Ferroic Materials - Understanding, Development, and Utilization

Edited by Ali Hussain and Maaz Khan

Published in London, United Kingdom

Ferroic Materials - Understanding, Development, and Utilization
http://dx.doi.org/10.5772/intechopen.1004969
Edited by Ali Hussain and Maaz Khan

Contributors
Ahmed Gadelmawla, Ali Hussain, Deepanshu Keneria, Fatima Rehan, Gwangseop Lee, Jung-Hyuk Koh, Liu Yang, Mohsin Saleem, Mudassar Shehzad, Muhammad Saad Ur Rahman, Muhammad Zubair Khan, Sergey G. Sandomirski, Shaoxiong Xie, Shedrach Yakubu, Shivam Kumar Mittal, Udeshwari Jamwal, Yaojin Wang

First published in London, United Kingdom, 2025 by IntechOpen
IntechOpen is the global imprint of INTECHOPEN LIMITED, registered in England and Wales, registration number: 11086078, 167-169 Great Portland Street, London, W1W 5PF, United Kingdom

For EU product safety concerns: IN TECH d.o.o., Prolaz Marije Krucifikse Kozulić 3, 51000 Rijeka, Croatia, info@intechopen.com or visit our website at intechopen.com.

British Library Cataloguing-in-Publication Data
A catalogue record for this book is available from the British Library

Ferroic Materials - Understanding, Development, and Utilization
Edited by Ali Hussain and Maaz Khan
p. cm.
Print ISBN 978-0-85014-972-2
Online ISBN 978-0-85014-971-5
eBook (PDF) ISBN 978-0-85014-973-9

Meet the editors

Dr. Ali Hussain is a Professor in the Department of Space Science at the Institute of Space Technology (IST) in Islamabad, Pakistan. He worked as an experienced researcher at the Department of Materials Science and Engineering, Friedrich Alexander University (FAU) Erlangen-Nuremberg, Germany, under the Alexander von Humboldt (AvH) Fellowship. He did his Ph.D. in Materials Science and Engineering at the University of Ulsan in South Korea and postdoctoral research at Changwon National University in South Korea. His research focuses on the development, characterization, and structural property relationship of environmentally benign lead-free ferroelectric materials for their applications in piezoelectric and energy storage devices. He has contributed to over 100 peer-reviewed research articles, 2 book chapters, and 1 registered patent.

Dr. Maaz Khan is the Deputy Chief Scientist at the Pakistan Institute of Nuclear Science and Technology, with a Ph.D. and postdoctoral expertise in materials science and nanoscience. His research specializes in the fabrication and comprehensive characterization of nanomaterials, including their structural, optical, magnetic, and electrical properties. He has authored over 150 research articles and 12 books. Currently, Dr. Khan is the editor-in-chief of *The Nucleus and the Journal of Materials, Processing, and Design* while also serving on the editorial boards of multiple other materials science journals.

Contents

Preface

Ferroic materials play a significant role in technology development and enhancement. These materials have unique physical properties and fascinating applications in the fields of electronics, healthcare, energy, telecommunications, transportation, etc. Ferroic materials are mainly categorized into three groups based on their ferroic orderings: ferroelectric, ferroelastic, and ferromagnetic. Understanding, developing, and utilizing ferroic materials is an active area of research. Academic, research, and industrial communities are focusing on exploring new ferroic materials, optimizing their properties for enhanced performance, novel functionalities, miniaturization, and integration in advanced technological devices. This book will provide the fundamental concepts of ferroic materials, polarization, magnetization, hysteresis, domain, domain wall motion, historical research evolution, synthesis, processing routes, and various characterizations. The structure, properties, relationships, and possible applications in various fields will be explored. This book will also discuss unresolved challenges and issues in this exciting and emerging field.

Ali Hussain
Institute of Space Technology (IST),
Islamabad, Pakistan

Department of Materials Science and Engineering,
Friedrich-Alexander-Universität Erlangen,
Nürnberg, Germany

Maaz Khan
Pakistan Institute of Nuclear Science and Technology,
Islamabad, Pakistan

Chapter 1

Ferroelasticity in Ferroic Materials

Shaoxiong Xie and Ali Hussain

Abstract

The macroscopic properties of ferroic materials are well-known to be influenced by their underlying domain dynamics. In ferroelectric ceramics, domain structures can be categorized into two main types: ferroelectric domains, associated with spontaneous polarization, and ferroelastic domains, linked to spontaneous strain. Currently, ferroelectric domains and their dynamic behaviors have been extensively studied, as they are recognized to directly influence the functionality of ferroelectric ceramics and determine the output performance of electronic devices. In contrast, the understanding of ferroelastic domain dynamics remains limited, primarily because ferroelastic deformation is only activated under mechanical loads of sufficient magnitude. Despite this, ferroelastic properties in ferroelectric ceramics must be considered in certain applications, such as piezoelectric multilayer actuators that are typically subjected to compressive preloads and high blocking forces. Under these loading conditions, ferroelastic deformation induces nonlinear behaviors and remanent strain, leading to degraded performance of ferroelectric ceramics and reduced stability and reliability of electronic devices. For this reason, deep understanding of ferroelasticity and deciphering mechanisms behind ferroelastic deformation in ferroelectric materials is essential for advancing material design and optimization to meet diverse application needs. This chapter explores the current understanding of ferroelastic properties and deformation mechanisms in representative ferroelectric materials under various loading conditions. Additionally, it highlights toughening effect related to ferroelasticity in ferroelectric materials that have been discovered in recent years. This knowledge could provide valuable insights for the development of the next generation of ferroic materials for a wide range of technological applications.

Keywords: ferroic materials, piezoelectrics and ferroelectrics, ferroelectric/ferroelastic domains, nonlinear behaviors, ferroelastic deformation, ferroelastic toughening, crack growth

1. Introduction

Ferroic materials have gained significant attention in both commercial and academic fields due to their wide-ranging applications in sensors, actuators, and energy storage. This widespread interest stems from their exceptional chemical and thermal stability, controllable multifunctionality, and remarkable mechanical properties [1–4]. Given their importance in our daily life and numerous practical

IntechOpen

advantages, there has been a growing focus on developing novel ferroic materials and advanced processing techniques. It is well known that the macroscopic behaviors of ferroic materials are fundamentally governed by their underlying domain dynamic behaviors [5]. The dynamics of these domains under applied fields play a critical role in determining the functional performance of materials, such as their piezoelectric, ferroelectric, dielectric, and magneto-electric properties. For instance, when an external electric field is applied to ferroelectric materials, it can induce 180°or non-180° domain switching, a phenomenon known as ferroelectric domain switching. This process involves the reorientation of the spontaneous polarization vector within the ferroelectric domains through a domain nucleation and growth process. In addition to electric fields, mechanical fields can also trigger domain switching events. However, unlike ferroelectric switching, domain switching induced by mechanical fields primarily involves non-180° domain changes, which are associated with the reorientation of the spontaneous strain within the domains [6]. This phenomenon, known as ferroelastic domain switching, occurs only under mechanical loads of sufficient magnitude along specific crystallographic orientations. The intriguing switching behaviors, encompassing both ferroelectric and ferroelastic domain dynamics, provide ferroelectric materials with significant potential to meet the demands of modern technological applications. Consequently, the optimization and fine-tuning of domain structures and their switching dynamics have become increasingly important research focuses in the field of ferroic materials today.

In practical engineered applications, ferroic materials are often simultaneously subjected to electrical, mechanical, and thermal loads [7]. Consequently, the performance, stability, and reliability of ferroic devices are significantly influenced by both their ferroelectric and ferroelastic properties under applied loads during operation. For instance, in current applications of piezoelectric multilayer actuators, these devices are typically exposed not only to electric fields but also to mechanical fields [8]. Mechanical fields, which often arise from compressive preloads and high blocking forces, frequently cause ferroelastic deformation and remanent strain, leading to reduced functional performance and significantly influencing the stability and reliability of ferroelectric devices during operation. In most application scenarios, even when the applied mechanical stress, such as preloading stress, is below the coercive stress, such moderate stress levels at high temperatures can still cause depolarization through ferroelastic deformation, thereby degrading the performance of ferroic devices [9]. However, attention to the mechanical stress-induced ferroelasticity is relatively scarce for ferroic materials, because ferroelasticity is not really considered a functional property that can be used for an application. In fact, it is simply an effect that occurs in ferroic materials during practical use. Ferroelectricity and ferromagnetism, on the other hand, are the primary functional response that people are using for their applications, and thus, much more attention has been paid to the electric/magnetic field-related functional behaviors of ferroic materials. Despite this, a comprehensive understanding of the ferroelasticity is essential for ferroic materials to ensure stable and reliable performance in practical service.

To investigate the ferroelasticity of ferroic materials, various testing methods have been employed to study their stress–strain nonlinear behaviors and ferroelastic dynamics, uncover the underlying mechanisms of ferroelasticity, and assess the impact of ferroelasticity on functional performance and mechanical properties [10–14]. Among these methods, the uniaxial compressive test is one of the most common approaches for characterizing ferroelastic behaviors, as ferroic materials can withstand compressive loads of sufficient magnitude. Three-point bending test is another effective technique for studying the ferroelasticity of ferroic materials, particularly in revealing crack growth and ferroelastic toughening behaviors. In addition to bending tests, indentation experiments are also used to examine crack growth and ferroelastic deformation-induced toughening effects. In certain cases, tensile tests are conducted to explore ferroelastic deformation in ferroic materials. However, because ferroic materials present intrinsic brittleness, tensile tests are usually performed using small-sized sample and analyzed using transmission electron microscopy (TEM) to enable *in-situ* observations of ferroelastic domain dynamics, providing insights into the underlying mechanisms of ferroelastic behaviors. Besides, a dynamic mechanical analyzer and other testing methods such as biaxial, triaxial, and shear tests can be also used to study ferroelasticity and reveal domain wall dynamics. Furthermore, advanced characterization techniques such as piezoresponse force microscopy (PFM), electron backscatter diffraction (EBSD), dual-beam focused ion beam (FIB), X-ray diffraction (XRD), neutron diffraction, Brillouin scattering, among others, are frequently employed. These methods offer new insights into ferroelasticity and ferroelastic dynamics, significantly enhancing our understanding of ferroelastic properties of ferroic materials.

Ferroelastic properties vary significantly across different materials, and their behaviors are highly dependent on testing conditions. Currently, some findings have already shown the experimental observations of ferroelasticity in ferroic materials [15, 16], while comprehensive understanding of ferroelastic deformation and its underlying mechanisms remains incomplete. Additionally, investigating ferroelasticity in certain novel ferrioc materials is challenging. For this reason, this section focuses on the findings related to ferroelasticity in several classical piezoelectric/ferroelectric materials that are one of the most important categories in ferroic materials, discussing the current understanding of ferroelastic behaviors under various loading conditions and revealing the effects of ferroelastic deformation on electrical performance and mechanical properties. The knowledge gained from studying ferroelasticity not only enhances our understanding of the ferroelastic properties of ferroic materials but also serves as a valuable reference for optimizing the performance of ferroic materials and the development of the next generation of ferroic materials for a wide range of applications.

2. General considerations of ferroelastic behaviors in ferroelectric materials

For ferroelectric materials, ferroelastic domains exhibit disordered spontaneous polarization distribution before mechanical loading. During compressive loading, ferroelectric materials first undergo a linear stress–strain response. This behavior occurs because ions shift only slightly from their equilibrium positions within the crystal structure, resulting in a linearly elastic response (**Figure 1**). As mechanical loads gradually increase, the stress–strain curve transitions to nonlinear behavior due to the onset of ferroelastic domain switching. At this stage, certain ions adopt new equilibrium positions, accompanied by the reorientation of ferroelastic domains to align perpendicular to the applied compressive load. With a further increase in mechanical load, the stress reaches coercive stress level, at which point ferroelastic domains tend to switch rapidly. This corresponds to the highest rate of domain switching, leading to a significant and rapid increase in remanent strain. As mechanical loads continue to increase, the material reaches a fully switched domain state or saturation. At this point,

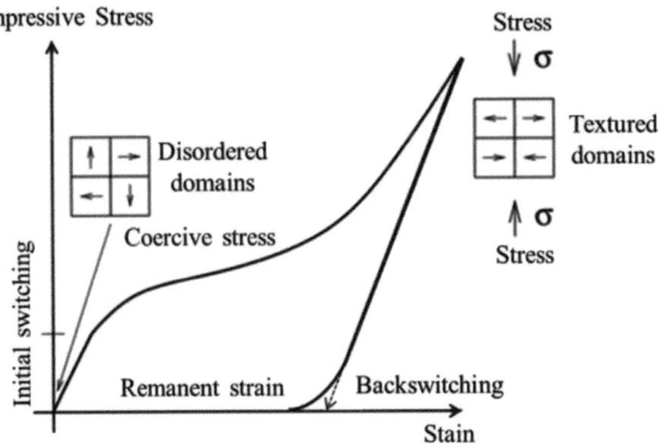

Figure 1.
Schematic of ferroelastic deformation under compressive loading.

all switchable ferroelastic domains have been reoriented, and the remanent strain reaches its maximum value. With further increases in mechanical load, the stress–strain curve returns to a linear response characterized by purely elastic deformation. During unloading, the stress–strain curve initially exhibits a linear response, followed by nonlinear deformation. This nonlinear behavior during unloading is attributed to ferroelastic domain backswitching, where local electrical and mechanical fields reorient a portion of the domains back to their original orientations, resulting in a reduction in the final remanent strain. The backswitching phenomenon is associated with reversible ferroelastic domains, while the remaining remanent strain primarily arises from irreversible ferroelastic domains. From this perspective, the ferroelastic switching is the primary mechanism responsible for the nonlinear deformation observed in ferroic materials under mechanical loading. These intriguing behaviors have been further validated using advanced techniques, such as XRD, PFM, TEM, and so on.

Generally, the ferroelastic properties are strongly influenced by various factors, including composition, domain structure, phase structure, temperature, poling state, and applied electric field, among others [16]. Additionally, loading rate and frequency can also significantly affect the ferroelastic response. In practical applications, greater attention must be given to the ferroelastic response, as excessive ferroelastic deformation caused by ferroelastic switching can lead to depolarization and increased remanent strain, thereby weakening the functional performance (such as piezoelectricity, ferroelectricity, elasticity, and dielectricity) of ferroelectric materials and reducing the reliability of ferroelectric devices. Understanding the ferroelastic properties of ferroelectric materials is therefore essential. This knowledge facilitates the optimization of their functional performance while improving the stability and reliability of ferroelectric devices. In the following section, several key factors influencing ferroelastic properties will be discussed.

3. Ferroelasticity in lead-based ferroelectric materials

Among ferroic materials, the ferroelastic properties of lead-based ferroelectric ceramics, such as lead zirconate titanate (PZT), have been extensively studied due to their exceptional electromechanical properties. PZT ceramics are generally categorized into two groups: soft PZT ceramics and hard PZT ceramics. Compared to soft PZT ceramics, hard PZT ceramics exhibit lower piezoelectric properties, higher coercive fields, smaller dielectric constants, reduced hysteretic loss, better aging characteristics, and superior mechanical stability [17]. These differences result in distinct ferroelastic switching and nonlinear deformation behaviors under applied mechanical loads for the two types of ceramics. Specifically, soft PZT ceramics, such as PIC 151 [11], display pronounced nonlinear ferroelastic deformation with a lower coercive stress. This indicates that ferroelastic switching occurs more readily in soft PZT ceramics, a characteristic attributed to their increased domain wall mobility. For hard PZT ceramics, such as PIC 181, they show a larger coercive stress, which contributes to their superior mechanical stability compared to the soft PZT ceramics. However, the hard PZT ceramics also display larger ferroelastic hysteresis and greater remanent strain, suggesting that numerous ferroelastic switching events occur during mechanical loading despite they higher coercive stress.

3.1 Ferroelectric phase-dependent ferroelastic properties of lead-based ferroelectric ceramics

The functional performance and mechanical properties of PZT ceramics can be tailored by phase boundaries, as different ferroelectric phases induce varying lattice distortions and domain structures, which in turn lead to distinct properties. For PZT ceramics with a tetragonal (T) ferroelectric phase, the larger crystal lattice distortion results in higher internal stress. This internal stress typically increases the energy barrier for ferroelastic switching, which explains why PZT ceramics with the T phase exhibit a higher coercive stress. However, PZT ceramics with multiphase coexistence also possess larger lattice distortion but lower coercive stress, which indicates that the lattice distortion is not the primary factor determining ferroelastic response [12]. Actually, for T phase, there exist 90° domains and 180° domains, while only 90° domains possess ferroelasticity. For rhombohedral (R) phase, 71° domains and 109° domains present ferroelasticity. For the PZT ceramics with R-T multiphase coexistence, they possess the 90° domains, 71° domains, and 109 ° domains with ferroelasticity. The presence of more non-180° domains in PZT ceramics with R-T multiphase coexistence enhances their properties, as the increased number of polarization orientations facilitates polarization switching. Additionally, the flattened free energy at the multiphase boundaries lowers the energy barrier for polarization reorientation, leading to improved properties. Therefore, the lower coercive stress in PZT ceramics with R-T multiphase coexistence could be primarily attributed to the higher number of switchable non-180° ferroelastic domains and the reduced free energy. Besides, the stress-induced phase transitions in PZT with multiphase coexistence have been revealed, which could be another mechanism for the change of coercive stress. Furthermore, compared to PZT ceramics with a single ferroelectric phase, those with R-T multiphase coexistence exhibit distinctly different ferroelastic behaviors. The initial slope of the loading curve significantly decreases for PZT ceramics with the R-T phase, resulting in a smaller elastic modulus. This makes PZT ceramics with multiphase coexistence more susceptible to mechanical loads during practical use, despite their excellent electrical properties.

3.2 Temperature-dependent ferroelastic properties of lead-based ferroelectric ceramics

Generally, PZT ceramics exhibit good temperature stability in terms of electrical performance. However, their ferroelastic properties are highly temperature-dependent. As the temperature increases, there is a noticeable reduction in coercive stress, remanent strain, and ferroelastic hysteresis [12]. When the applied temperature exceeds Curie temperature, the stress–strain curve shows a purely linear elastic response, similar to the linear stiffness behavior observed at lower temperatures [9], which is attributed to the diminished lattice distortion due to the nature of non-ferroelastic paraelectric phase over Curie temperature. The decrease in coercive stress at high temperatures is due to the fact that ferroelastic domains are more easily switched because of the lower energy barrier during heating. However, as heating progresses, the gradual degradation of ferroelastic domains reduces the number of switchable domains, which corresponds to a reduction in spontaneous strain. This also leads to smaller remanent strain and lower ferroelastic hysteresis. Interestingly, PZT ceramics with multiphase coexistence exhibit better temperature stability in their ferroelastic response, as evidenced by minimal variation in coercive stress. This suggests that ferroic materials with multiphase boundary design are expected to have improved temperature resistance to ferroelastic deformation during service.

3.3 Poling states-dependent ferroelastic properties of lead-based ferroelectric ceramics

Poling states play significant role in influencing ferroelastic response of PZT ceramics. Actually, for the PZT ceramics used in practical applications, they are poled under applied electric field, thus showing piezoelectric properties or other functional performance. Therefore, the understanding of ferroelastic properties of poled PZT ceramics is critical for us to develop this kind of ferroic materials. For the PZT ceramics with poling treatment, they also exhibit representative ferroelastic behaviors under compression loading, i.e., the stress–strain curve shows nonlinear deformation and linear stiffness behaviors (**Figure 2a**). Compared to unpoled PZT ceramics, the poled PZT ceramics exhibit increased nonlinearity and higher remanent strain (**Figure 2a, b**). This is because numerous ferroelastic domains will be aligned in the direction of mechanical loading after poling process, contributing to more ferroelastic switching events [9]. Furthermore, the ferroelastic behavior of poled PZT ceramics is highly temperature-sensitive, with variations in nonlinear deformation and remanent strain at different temperatures being greater than those observed in unpoled PZT ceramics. Additionally, the coercive stress in poled PZT ceramics is

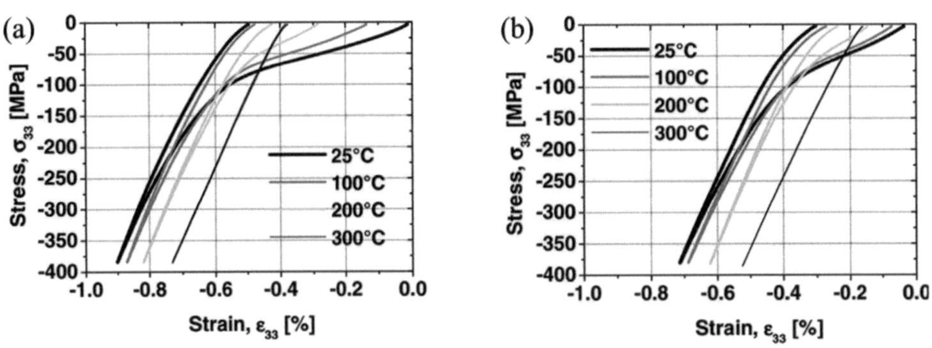

Figure 2.
Stress–strain response at different temperatures for compressive mechanical loading of (a) poled and (b) unpoled PZT samples [9].

lower than in unpoled PZT ceramics, and temperature further reduces coercive stress in poled ceramics. This highlights the importance of considering the ferroelastic properties of PZT ceramics in practical applications, especially at high temperatures, where even small mechanical loads can cause performance degradation and reduced stability in ferroelectric devices.

3.4 Loading rates-dependent ferroelastic properties of lead-based ferroelectric ceramics

Loading rates are also an importance consideration when we analyze the ferroelastic properties of PZT ceramics. For the PZT ceramics, strain response depends strongly on the loading rates, especially in the nonlinear deformation region [18]. Specifically, fully ferroelastic switching can be induced at lower loading rates, leading to a higher amount of induced strain. This suggests that a final stable irreversible strain state could be realized at very low loading rate. However, at higher loading rates, the switchable ferroelastic domains cannot fully reorient due to insufficient loading time, thereby resulting in relatively smaller remanent strain. To reduce the influence of loading rates related time-dependent effects, pulse-shaped partial unloading method can be performed during the mechanical loading. Partially unloading ceramics at an appropriate stress correspond to linear and non-hysteretic behaviors, and thus, the stress–strain response could be not influenced by partial unloading [11]. Interestingly, partial unloading provides sufficient time for all switchable ferroelastic domains to reorient under mechanical loading, particularly at lower loading rates. This allows for more ferroelastic switching events, resulting in a larger remanent strain that is closer to its true value.

3.5 Ferroelastic response under applied electric bias fields

In practical use, PZT ceramics are generally serviced under applied electric bias field. The various applied fields can induce distinct ferroelastic responses, while the coupling effect of mechanical and electric fields makes the ferroelastic behaviors more complicated. For example, the ferroelastic deformation of poled PZT ceramics can be significantly suppressed by applying a negative bias electric field aligned with the direction of the previous poling [19]. As the magnitude of the negative bias electric field increases, ferroelastic switching becomes increasingly difficult (**Figure 3a**). In this case, higher mechanical stresses are required to initiate the mechanically induced ferroelastic switching events, resulting to a larger coercive stress. This intriguing behaviors have been also revealed elsewhere [10]. Furthermore, the back-switching behaviors are also profound at larger negative bias electric field, indicating significant polarization recovery processes of ferroelastic domains (**Figure 3b**). However, when a positive bias electric field with orientation opposite to the poling direction is applied to PZT ceramics, the ferroelastic deformation becomes easier. This is because the positive bias electric field can induce additional domain switching in the transverse direction in addition to stress-induced ferroelastic switching. The phenomenon could be explained using following equation [20, 21]:

$$\sigma_{ij}\Delta\varepsilon_{ij} + E_i\Delta P_i \geq G_c \tag{1}$$

where E_i and ΔP_i represent bias electric field and change in spontaneous polarization; σ_{ij} is applied stress, and $\Delta\varepsilon_{ij}$ is the change in spontaneous strain; G_c is a critical

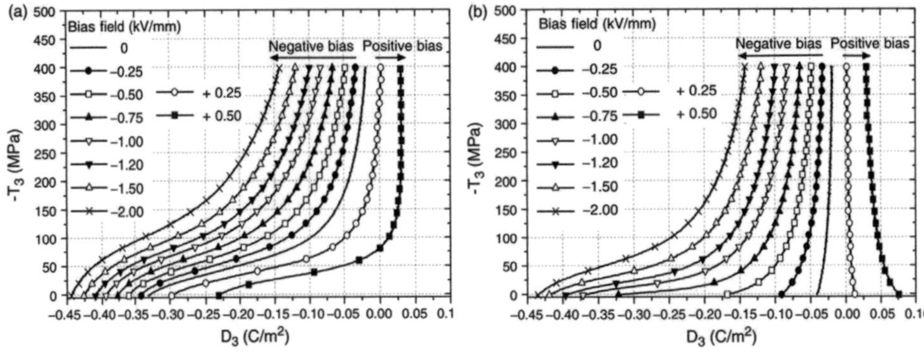

Figure 3.
Stress–strain curves of PZT ceramics as a function of bias electric field. (a) Mechanical loading processes and (b) unloading processes at different bias electric fields [19].

threshold value for switching events to take place. From the polarization switching criterion, at certain a stress, bias electric field can significantly affect the critical threshold value for ferroelastic switching, thereby determining the final ferroelastic behaviors. Obviously, the positive bias could promote the ferroelastic deformation, while negative bias tends to hinder the ferroelastic deformation. Interestingly, PZT ceramics exhibit a higher final remanent strain with the assistance of a positive bias. After unloading, the positive bias induces some domains to orient in the direction of the bias. In other words, the transition from loading to unloading in PZT ceramics under positive bias involves a process of depolarization followed by polarization, resulting in a remarkable polarization value after one loading cycle. It is important to note that the ferroelastic deformation of ferroic materials becomes more complicated when they are subjected to the combined effects of temperature and electric bias fields. Temperature can promote ferroelastic deformation, while the electric bias field can influence the ferroelastic response. Consequently, the interaction between the temperature and electric bias fields leads to varying ferroelastic properties, making ferroic materials more sensitive to applied mechanical loads [22].

3.6 Ferroelastic responses and fatigue behaviors under cyclic loading

Another interesting aspect of PZT ceramics is their ferroelastic response under cyclic loading. The ferroelastic deformation induced by cyclic loads is influenced by various factors, such as stress levels, loading rate, electrode status, microstructure, and the like. Generally, remanent or irreversible strain serves as an indicator of the evolution of ferroelastic behavior under cyclic loading, as they respond sensitively to cyclic loads during ferroelastic switching. Specifically, higher stress amplitudes and slower loading rates during cyclic loading tend to induce more irreversible strain [23]. This is because ferroelastic domains switch more rapidly and are more likely to reach a saturated state at higher stress levels, while at slower loading rates, the switchable ferroelastic domains can fully orient due to the time-dependent effect. For the ferroelastic behaviors under cyclic loading, the most notable observation is that the irreversible strain decreases as the number of cycles increases, with the majority of irreversible strain occurring in the first few cycles [24]. This indicates the ferroelastic deformation gradually saturates with the increase of the number of cyclic loading, and the ferroelastic switching primarily occurs in the early stages of cyclic loading.

The cyclic loading generally causes fatigue response in materials. Similar with metals, stress amplitude versus cycles to failure curve under three point bending testing could be obtained for PZT ceramics. With the increase of cycles to failure, the stress amplitude decreases gradually, exhibiting a common fatigue response under cyclic loads. However, different electrodes of ferroic materials show various fatigue resistances. For instance, fatigue strength for nickel electrode is larger than that for silver electrode [25]. This is because nickel electrode has less porous layer than silver electrode, and the surface roughness is also better for the former. Besides this, the fatigue behaviors of ferroic materials are related to the feroelastic switching process, where ferroelastic deformation makes the materials harder, and the hardening behaviors are resulting from the ferroelastic switching. Therefore, for ferroic materials, the hardening is also a hint of ferroelastic response and fatigue behaviors during cyclic loading.

3.7 *In-situ* characterizations of ferroelastic deformation

In addition to the macroscopic observations of ferroelastic behaviors of PZT-based ferroic materials, *in-situ* characterizations at micro-scale are also crucial for understanding ferroelastic deformation, which can be performed using various advanced techniques. The most common approach to characterize the ferroelastic switching is XRD analysis. According to the XRD diffraction evolution of representative twin-related peaks, such as (200) and (002), quantitative comparison between the degree of domain orientation and nonlinear properties of PZT ceramics could be analyzed, thereby revealing the corresponding ferroelastic deformation during mechanical loading [26]. Neutron diffraction can be also used for the studies of ferroelasticity of ferroelectric ceramics, and the corresponding analysis of ferroelectric switching is similar with the XRD characterization. PFM is a powerful tool to characterize domain structures in ferroic materials based on the inverse piezoelectric effect [5]. Therefore, *in-situ* observations of ferroelastic domain switching for PZT materials could be well carried out by PFM technique. For example, the PFM images could be obtained during each mechanical loading stage, and the ferroelastic domain structures are clearly revealed from these PFM images (**Figure 4**) [21]. Before mechanical loading, the ferroelastic domains in PZT ceramics exhibit diverse distributions and crystallographic orientations, with representative domain structures highlighted in the PFM images (**Figure 4a**). After mechanical loading, these domains undergo significant evolution, characterized by domain broadening, domain wall motion, and domain overwriting, which serve as primary indicators of mechanically induced ferroelastic deformation (**Figure 4b,c**). During ferroelastic switching, particularly in the domain overwriting process, multiple switching events occur, including both energetically favorable and unfavorable ferroelastic switching. Notably, ferroelastic deformation can be also influenced by the pre-existing domain structures, particularly in polycrystalline ceramic materials. Upon unloading, some switched ferroelastic domains partially revert to their original states (**Figure 4d**). The corresponding schematics of the ferroelastic switching processes, shown in **Figure 4e,f**, illustrate the intriguing domain evolutions during mechanical loading.

In addition to PFM analysis, EBSD technique is also a viable approach for characterizing ferroelastic responses in ferroelectric materials based on the different orientation reflections of ferroelastic domains. For PZT ceramics, EBSD results reveal that ferroelastic domain switching occurs within the range of 85.4 to 90°, with domains crossing to the opposite side of the compressive loading axis [27]. A ferroelastic

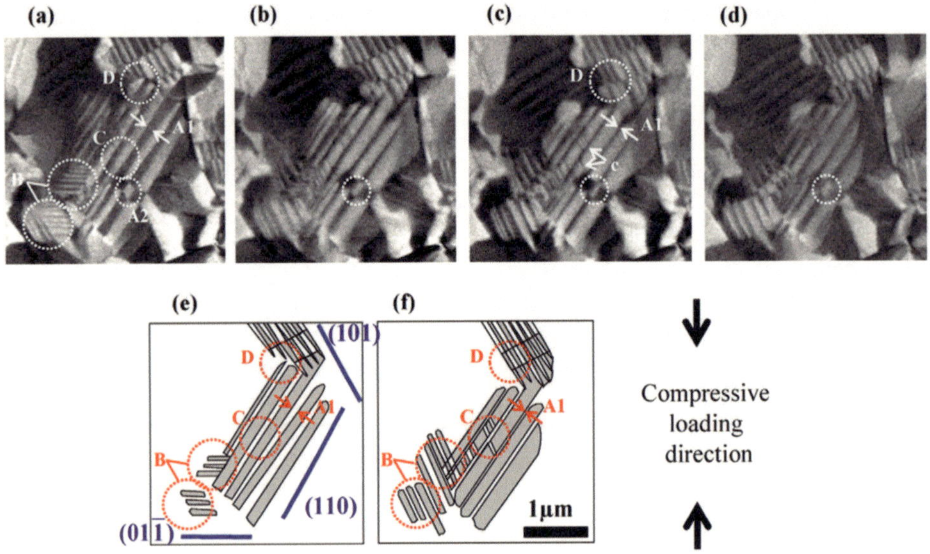

Figure 4.
PFM images (a–d) show domain evolution processes under compressive loading. Two schematics (e–f) depict key evolution processes before loading and after loading [21]. (a) Before loading (b) A 0.118% strain (c) A 0.134% strain (d) After unloading (e) Before loading (f) A 0.134% strain.

switching model has been proposed based on the twin deformation model, where the switching angle is closely related to the aspect ratio of crystal parameters. Additionally, the structural analysis of ferroelastic deformation in PZT materials can be conducted using the TEM technique. Investigations of the ferroelastic properties of PZT materials have shown that ferroelastic switching is most readily observed at highly active needle points within ferroelastic domains. This switching can be stabilized and pinned by dislocations, leading to an incomplete switching process [28]. Such ferroelastic switching process is also validated by phase-field modeling, which provides new insights into fundamental physics of ferroelastic deformation in ferroic materials.

4. Ferroelastic properties of lead-free ferroelectric materials

Barium titanate (BT) is a type of representative lead-free ferroelectric materials used in many applications. The stress–strain responses of BT ceramics are similar with that of lead-based ceramics, i.e., they also undergo first elastic linear deformation and then ferroelastic nonlinear deformation during mechanical loading. The understanding of ferroelastic properties of BT ceramics primarily focuses on the experimental observations of ferroelastic domain evolutions during mechanical loading. For example, *in-situ* neutron diffraction results indicate that under maximum load, 12% of ferroelastic domains undergo switching. Among these, 5% revert to their original states upon unloading, while only 7% remain in their new equilibrium polarization states [29]. PFM observations reveal that the nucleation of new ferroelastic domains in BT ferroelectric materials can be induced by the local accumulation of surface charges. Antiparallel polarized a-domains may be separated by charged 180° domain walls, which evolve in morphology and give rise to fascinating ferroelastic deformation behaviors under compressive stress [30]. In addition to BT materials, potassium sodium niobate (KNN) ceramics are another type of representative lead-free ferroelectric materials, which also present excellent electro-mechanical properties and show huge potentials to replace lead-based PZT materials in future. For this reason, the ferroelastic properties of KNN materials have also been investigated [31]. The resulting findings are also consistent with that in lead-based materials. However, the understanding of ferroelastic domains and related deformation mechanisms of KNN lead-free materials is still scarce. The knowledge of ferroelastic properties of lead-based materials could provide constructive information for further understanding nonlinear behaviors of KNN materials.

5. Ferroelastic properties of high-temperature ferroelectric materials

Bismuth titanate (BIT), a representative compound among high-temperature bismuth layer structured ferroelectrics (BLSFs), exhibits classic nonlinear deformation behaviors in the stress–strain curve during mechanical loading. The nonlinear deformation is generally attributed to the pseudo-90° and pseudo-180° ferroelastic switching behaviors [6, 32–34]. The BIT ceramics with higher sintering temperatures display better ferroelastic switching behavior but lower failure stress. The ferroelastic responses under multi-field conditions have also studied for another representative high-temperature BLSFs, calcium bismuth titanate (CBT) [22]. Specially, poling could increase remanent strain and decrease elastic modulus. Elevated temperature could increase energy dissipation and improve ferroelastic domain wall mobility. Transverse electric field could increase remanent strain and cause complicated fracture behaviors. For these reasons, multi-field conditions significantly affect ferroelastic properties and fracture behaviors, while the ferroelastic deformation of CBT materials with higher Nb/Mn doping levels present better resistance to multi-field coupling effects.

6. Ferroelastic properties of relaxor ferroelectric materials

Relaxor ferroelectric materials are also representative advanced functional materials, which have been widely investigated for various applications. However, different from normal ferroelectric materials, the domains inside relaxor ferroelectric materials are not long-range polarization order but short-range ordered polar nano-regions [35, 36], which lead to a various ferroelastic deformation response in ferroelectric materials under mechanical loading. For example, under compressive loading, NBT-based relaxor ferroelectric ceramics present an obvious nonlinear stress–strain response, which is attributed to the domain evolution from short-range ordered polar nano-regions to long-range ordered domain structures [37]. It is worth noting that the NBT-based relaxor ferroelectric ceramics with multi-phase coexistence usually present higher coercive stress, which is different from the case in normal ferroelectric ceramics, such as PZT and BT. This is because that the nonlinear deformation in relaxor ferroelectrics is resulted from the formation of long-range ordered domains, whereas the mechanically induced domain evolution process needs to overcome more energy barrier compared with stress-induced domain wall motion in normal ferroelectric ceramics, thus showing a larger coercive stress. After electrical poling treatment of NBT-based relaxor ferroelectric ceramics, the long-range ordered domain structures are generated under applied electric field, whose orientations align to the direction paralleling to the mechanical loading, thus leading to a more ferroelastic switching events and producing more remanent strain under mechanical loading. Additionally, the coercive stress in poled relaxor ferroelectric ceramics is smaller, with this case similar with that in normal ferroelectric ceramics. Additionally, composition-dependent ferroelastic response is also obvious in relaxor ferroelectric materials. For instance, the NBT-based relaxor ferroelectric ceramics with the addition of different KNN contents, their nonlinear ferroelastic properties exhibit significant changes [16]. Specifically, the coercive stress increases but remanent strain decreases for the ceramics with higher KNN contents. Interestingly, the backswitching behaviors tend to increase in relaxor ceramics, which are likely attributed to a reversible phase transition from ferroelectric phase to relaxor phase upon unloading, and the relaxation behaviors facilitate such backswitching process.

7. Ferroelastic properties of antiferroelectric materials

With growing concerns about energy crisis, antiferroelectric materials have garnered significant interest for energy storage technologies due to their exceptional energy storage density and rapid charge–discharge capabilities. A particularly intriguing characteristic of antiferroelectric materials, such as sodium niobate (NN) and silver niobate (AN) ceramics, is their complex structural phase transitions at different temperatures. These transitions result in various phase structures observed at ambient conditions, giving rise to fascinating ferroelastic properties. For instance, AN ceramics display distinct ferroelastic hysteresis and remnant strain behaviors in the weak ferroelectric, antiferroelectric, and paraelectric phase regions, respectively. These behaviors are primarily attributed to their differences in crystal symmetry, lattice distortion, thermal barriers to ferroelastic switching, and ferroelastic deformation responses under uniaxial compressive stress [38]. The different degrees of remanent domain textures and lattice strains have been observed in AN ceramics, with their magnitudes significantly influenced by the different phase states under applied stress. Additionally, *in-situ* stress Raman spectroscopy has detected changes in oxygen octahedral configurations and cation displacements during mechanical loading in AN ceramics. These structural evolutions are likely linked to ferroelastic responses, particularly in high-temperature phases, which can induce notable changes in these structural characteristics. These insights into antiferroelectric materials contribute to optimizing the functional performance of energy storage systems by tailoring mechanically induced ferroelastic responses. However, the current understanding of ferroelasticity in antiferroelectric materials remains limited, highlighting the need for further investigations in the future.

8. Ferroelastic properties of organic–inorganic hybrid ferroelectric materials

As emerging class of advanced materials, organic–inorganic hybrid ferroelectric materials have been extensively investigated due to their time-efficiency synthesis, ease of production, and low cost. It is well-known that domain structures play significant role in tailoring photoelectrical or photovoltaic properties, sparking significant research interest in understanding ferroelasticity in these materials. For example,

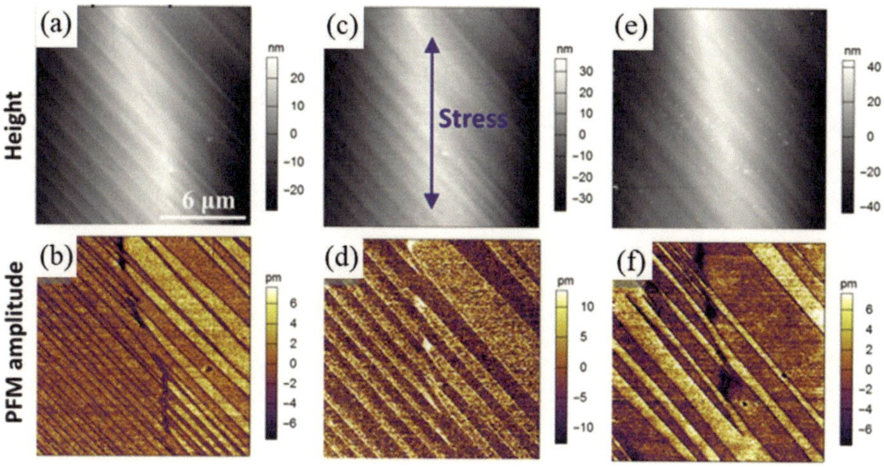

Figure 5.
PFM observations of ferroelastic domains modulated by external stress. Topography and PFM amplitude images of $CH_3NH_3PbI_3$ materials with (a, b) pristine state, (c, d) mechanical loading, and (e, f) relieving the stress, respectively [39].

the ferroelastic properties have been revealed in a representative organic–inorganic hybrid ferroelectric material, $CH_3NH_3PbI_3$ ($MAPbI_3$). The striped ferroelastic domains have been observed in $MAPbI_3$ materials, while these domains present strong contrast in the PFM amplitude images but weak contrast in the PFM phase images, significantly indicating the nature of ferroelasticity [39]. To figure out the stress-induced ferroelastic dynamics, bending stress is applied in $MAPbI_3$ materials (**Figure 5**). **Figure 5(a,b)** shows the ferroelastic domains before mechanical loading, showing striped ferroelastic domain features. After bending stress (**Figure 5c,d**), the narrow striped ferroelastic domains get together to form wider stripes, and dark purple ferroelastic domains could not be observed any more, strongly reflecting an obvious ferroelastic switching process. Upon unloading (**Figure 5e,f**), the ferroelastic domains could not go back to their original state, while the ratio between bright and dark ferroelastic domains significantly evolves, revealing mechanically-induced irreversible ferroelastic deformation. The findings of ferroelasticity in organic–inorganic hybrid ferroelectric materials are also similar with that observed in traditional lead-based or lead-free ferroelectric materials. Deep studies on ferroelastic properties of novel organic–inorganic hybrid ferroelectric materials are required in future, which could provide constructive values for further optimizing functional performance of advanced photoelectrical or photovoltaic devices.

9. Size-dependent ferroelastic properties

As we know, under applied mechanical stress, ferroelectric materials exhibit nonlinear stress–strain responses and produce remanent strain due to ferroelastic deformation. However, when sample size reduces to 2 µm (such as a micro-pillar sample), ferroelectric materials do not produce remanent strain, even though the ferroelastic deformation is observed under mechanical loading (**Figure 6**). In this case, ferroelectric materials behave as superelastic materials [40]. The intriguing behaviors are related to the surface tension of the micro-pillar, where surface tension modulates 90°

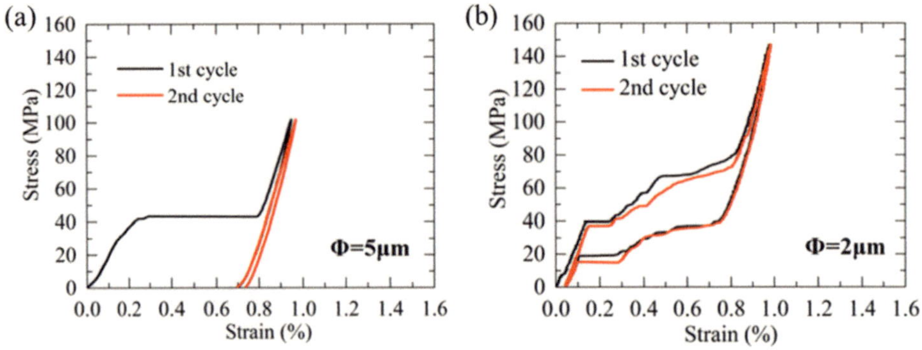

Figure 6.
Stress–strain response of BT ceramics with different sample sizes. (a) 5 μm and (b) 2 μm [40].

ferroelastic domain switching, responsible for the superelastic behavior. Specifically, when the sample size is larger, the surface tension is smaller, making it more difficult to induce ferroelastic domains to return to their original state after mechanical loading, thus resulting in remanent strain. However, as the sample size reduces to a critical value, the surface tension increases significantly. This larger surface tension helps switched ferroelastic domains return to their previous polarization states after mechanical loading, leading to superelastic behavior without remanent strain.

Additionally, the small-sized ferroelectric materials exhibit excellent fatigue resistance properties, which are attributed to the small energy barrier and ultralow energy dissipation during mechanical loading. The surface tension-modulated ferroelastic deformation has not been observed in macroscopic studies of ferroelectric materials. Furthermore, superelastic behaviors have only been observed in a limited range of ferroelectric materials. Further research on size-dependent ferroelastic deformation and superelastic behavior in various ferroelectric materials is needed. This knowledge could not only offer novel strategies for the design and optimization of dampers in micro-electromechanical systems (MEMS) devices, but also provide valuable insights into addressing fatigue issues in a wide range of potential applications.

10. 180° ferroelectric domain switching during ferroelastic deformation

Generally, ferroelastic deformation in ferroelectric materials is only related to non-180° ferroelastic domains, since the spontaneous strain could be only generated from non-180° ferroelastic domain switching process. For 180° domains, they are ferroelectric and will not generate spontaneous strain during applied mechanical loading, as their polarization vectors are of the same magnitude but opposite directions. In most ferroelastic deformation events, the applied stress is relatively low, as in the case of piezoelectric ceramics, since higher stress levels can lead to cracks or even fracture. At lower stress levels, ferroelastic deformation primarily involves non-180° ferroelastic switching, with both reversible strain (due to back-switching) and irreversible strain (due to ferroelastic deformation) occurring. However, higher stress levels could be applied to small-sized ferroelectric materials during *in-situ* TEM characterization. Experimental results indicate that energy barrier for ferroelastic switching in small-sized sample is significantly lowered, leading to the formation of a single domain that does not revert after unloading, accompanied by the occurrence of 180° ferroelectric switching [41]. Despite the 180° ferroelectric switching does not directly contribute to ferroelastic deformation, it can enhance the remanent polarization of ferroic materials, which is beneficial for non-volatile memory and other ferroelectric applications. This suggests that mechanical stress can play a similar role as an electric field in manipulating ferroelectric domain switching and influencing electrical performance when the applied stress is large enough. Research on mechanically-induced ferroelectric switching is still in its early stages, with many aspects yet to be explored in the future.

11. Ferroelastic toughening

11.1 Indentation crack growth behaviors

Ferroelastic domains can be reoriented under applied stress, and such ferroelastic switching behaviors play significant role in regulating crack growth behaviors. In particular, ferroelastic deformation exhibits a toughening effect, enhancing the material's resistance to crack growth. The phenomenon of ferroelastic toughening could be observed in indentation measurements for ferroelectric materials. Under Vickers indenter, there will be a symmetric rhombic indentation induced on the surface of piezoelectric ceramics. Due to releasing strain energy produced by elastic deformation inside indentation, indentation cracks are generated from rhombic angles and grow along the diagonal direction of rhombic indentation [13]. It is noteworthy that the length of the resulting cracks remains nearly identical, even when crack deflection or branching occurs in some instances. However, when an electric field is applied during indentation measurements of piezoelectric ceramics, significant anisotropic crack growth behavior is observed [14]. Specifically, when the electric field aligns with the direction of crack growth, the crack growth is hindered, resulting in shorter crack lengths. Conversely, cracks oriented perpendicular to the electric field exhibit greater growth lengths. This intriguing phenomenon is illustrated in **Figure 7** [42]. Before electric poling, indentation cracks demonstrate isotropic growth (**Figure 7a**). After applying electric poling field, anisotropic crack growth behaviors emerge (**Figure 7b**), with the growth rate of cracks parallel to the electric field being significantly slower than that of cracks perpendicular to it.

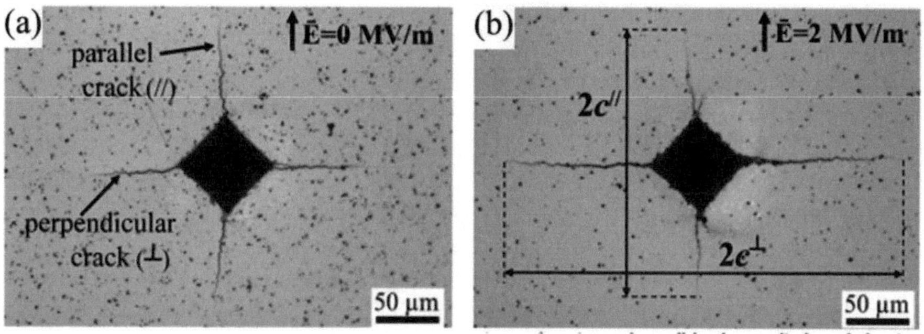

Figure 7.
Indentation cracks parallel and perpendicular to the poling direction. (a) unpoled sample and (b) poled sample [42].

This anisotropy is attributed to the ferroelastic toughening effect, driven by ferroelastic switching events occurring near the crack tips. Specifically, the material exhibits higher resistance to indentation crack growth along the direction parallel to the applied electric field, indicating the presence of numerous ferroelastic switching events and a stronger ferroelastic toughening effect in this orientation.

11.2 Pre-crack growth behaviors

In addition to indentation experiments, the ferroelastic toughening behavior of ferroic materials is often evaluated using the single-edge-notch-beam (SENB) method under bending loads. In SENB testing, samples are prepared as cuboids with special longitudinal, height, and thickness dimensions and then poled along three different orientations. During bending tests, crack growth initiates at the pre-crack region. As the crack propagates, a process zone forms, and the height of the process zone, h, is determined using the following equation [43]:

$$h = B\left(\frac{K_{IR}}{\sigma_{th}}\right)^2 \tag{2}$$

where $B = 1/2\pi$ is scaling factor for plane stress state; K_{IR} is total mode I crack resistance; σ_{th} is threshold stress to induce ferroelastic switching process. The process zone is generated due to the ferroelastic switching, and thus, the height of process zone could determine the ability of ferroelastic toughening of ferroelectric materials. For example, in ferroelectric ceramics poled along the longitudinal direction (where the domains align perpendicular to the pre-crack), the height of the process zone is smaller compared to ceramics poled along the height direction (where the domains align parallel to the pre-crack). As a result, the ceramics poled along the height direction exhibit higher fracture toughness due to a stronger ferroelastic toughening effect. Interestingly, when the ceramics are poled along the thickness direction, the crack tip stress can induce ferroelastic switching in various orientations, leading to a significant ferroelastic toughening effect and enhanced fracture toughness [44].

11.3 R-curve behavior

Another fascinating research topic related to ferroelastic toughening is R-curve behavior of ferroelectric materials [45], as R-curve behavior plays crucial roles in optimizing anti-cracking design and predicting fracture failure of smart devices during practical engineering applications. R-curve describes fracture toughness as a function of crack growth, reflecting the ability of crack growth and fracture properties of materials. In a crack growth event, fracture intensity factor of crack tip is determined by both intrinsic fracture toughness and extrinsic toughening effects, such as ferroelastic toughening or other microstructural aspects. The R-curve usually exhibits a nonlinear change with the crack growth, where the R-curve behavior is found to exhibit increasing fracture toughness during crack growth, and the change rate of R-curve reflects the crack growth resistance resulting from ferroelastic toughening (**Figure 8a**) [46]. The rapid nonlinear increase of R-curve corresponds to the larger ferroelastic toughening effect, and as a result, the ferroelectric materials will gain a higher fracture intensity factor of crack tip and present stronger crack growth resistance.

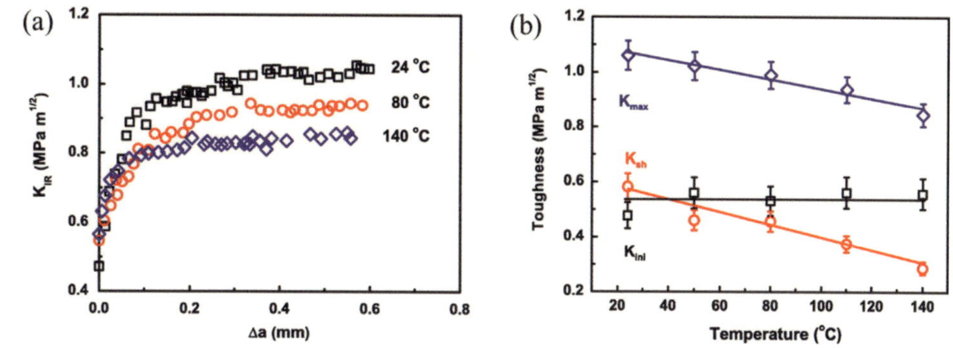

Figure 8.
R-curve behavior of PZT ferroelectric ceramics. (a) Temperature-dependent R-curve behavior at three different temperatures. (b) Maximum, shielding, and initial toughness values extracted from R-curves measured at elevated temperature [46].

Additionally, R-curve behavior is also influenced by grain size, domain orientation, doping element, poling state, ambient temperature, and the like. These factors could alter ferroelastic behaviors and thus affect fracture properties of ferroelectric materials. For instance, there is a decrease in the maximum toughness (K_{max}) and shielding toughness (K_{sh}) with the increase of temperature, while the initial toughness (K_{ini}) keeps only subtle change, suggesting the decreased toughening effect at high temperature (**Figure 8b**). Currently, the R-curve response has been widely studied in lead-based ferroic materials, while related research in lead-free materials is scarce. As the demand for high-reliability and high-stability electronic devices grows, it is essential to gain a deeper understanding of the R-curve behavior in lead-free ferroelectric materials in the future.

12. Conclusion

This chapter discussed the general understanding of ferroelastic behavior in ferroelectric materials and explored temperature, poling state, loading rate, bias field, and cyclic loading-dependent ferroelastic properties and related deformation mechanisms in lead-based ferroelectric ceramics. Besides this, the current findings of ferroelasticity in other representative ferroelectric materials, including lead-free ceramics, high-temperature ceramics, relaxor ceramics, antiferroelectric ceramics, and organic–inorganic hybrid ferroelectrics, have been discussed. Moreover, we introduced ferroelastic toughening behavior and its effect on fracture mechanics of ferroelectric materials. This knowledge contributes to the material design and performance optimization of novel ferroic materials for various applications.

Acknowledgements

Shaoxiong Xie thanks the Alexander von Humboldt Foundation for its fellowship support.

Author details

Shaoxiong Xie* and Ali Hussain
Department of Materials Science and Engineering, Friedrich-Alexander Universität Erlangen-Nürnberg, Erlangen, Germany

*Address all correspondence to: xsxdyx@126.com

References

[1] Luo X, Yan Z, Luo H, Zhou X, Li B, Zhang M, et al. Greatly improved piezoelectricity and thermal stability of (Na,Sm) Co-doped $CaBi_2Nb_2O_9$ ceramics. Advanced Powder Materials. 2023;**2**:100116

[2] Xie S, Xu Q, Chen Q, Zhu J, Wang Q. Realizing super-high piezoelectricity and excellent fatigue resistance in domain-engineered bismuth titanate ferroelectrics. Advanced Functional Materials. 2024;**34**:2312645

[3] Zhang S. High entropy design: A new pathway to promote the piezoelectricity and dielectric energy storage in perovskite oxides. Microstructures. 2023;**3**:202300

[4] Li J, Zhang Y, Yan M, Zhong C, Zhao L, Zhai D, et al. 3D printing of flexible piezoelectric composite with integrated sensing and actuation applications. Advanced Powder Materials. 2024;**3**:100226

[5] Xie S, Chen Y, Liu W, Xia G, Huang B, Liu C, et al. Three-dimensional domain patterns in tetragonal-to-monoclinic $Bi_4Ti_3O_{12}$ ceramics: Nonlinear analysis and piezoresponse force microscopy imaging. Acta Materialia. 2020;**188**:228-240

[6] Xie S, Tan Z, Jiang L, Nie R, Xu Q, Chen Y, et al. Ferroelastic properties and compressive stress-strain response of bismuth titanate based ferroelectrics. Ceramics International. 2020;**46**:1183-1188

[7] Xie S, Xu J, Chen Y, Tan Z, Nie R, Wang Q, et al. Flexural fracture mechanisms and fatigue behaviors of $Bi_4Ti_3O_{12}$-based high-temperature piezoceramics sintered at different temperatures. Ceramics International. 2018;**44**:16758-16765

[8] Randall C, Kelnberger A, Yang G, Eitel R, Shrout T. High strain piezoelectric multilayer actuators-a material science and engineering challenge. Journal of Electroceramics. 2005;**14**:177-191

[9] Webber K, Aulbach E, Key T, Marsilius M, Granzow T, Rödel J. Temperature-dependent ferroelastic switching of soft lead zirconate titanate. Acta Materialia. 2009;**57**:4614-4623

[10] Schäufele AB, Heinz HK. Ferroelastic properties of lead zirconate titanate ceramics. Journal of the American Ceramic Society. 1996;**79**:2637-2640

[11] Marsilius M, Webber KG, Aulbach E, Granzow T. Comparison of the temperature-dependent ferroelastic behavior of hard and soft lead zirconate titanate ceramics. Journal of the American Ceramic Society. 2010;**93**:2850-2856

[12] Seo Y-H, Franzbach DJ, Koruza J, Benčan A, Malič B, Kosec M, et al. Nonlinear stress-strain behavior and stress-induced phase transitions in soft $Pb(Zr_{1-x}Ti_x)O_3$ at the morphotropic phase boundary. Physical Review B-Condensed Matter and Materials Physics. 2013;**87**:094116

[13] Xie S, Xu J, Chen Y, Tan Z, Nie R, Wang Q, et al. Indentation behavior and mechanical properties of tungsten/chromium co-doped bismuth titanate ceramics sintered at different temperatures. Materials. 2018;**11**:503

[14] Xu Q, Xie S, Zhao F, Li L, Zhang Y, Zhu J, et al. Fracture behaviors and crack

propagation anisotropy in multi-ions modified bismuth titanate ferroelectrics. Journal of Alloys and Compounds. 2024;**1009**:176880

[15] Kamlah M. Ferroelectric and ferroelastic piezoceramics modeling of electromechanical hysteresis phenomena. Continuum Mechanics and Thermodynamics. 2001;**4**:219-268

[16] Webber KG, Vögler M, Khansur NH, Kaeswurm B, Daniels JE, Schader FH. Review of the mechanical and fracture behavior of perovskite lead-free ferroelectrics for actuator applications. Smart Materials and Structures. 2017;**26**:063001

[17] Berlincourt D. Piezoelectric ceramic compositional development. The Journal of the Acoustical Society of America. 1992;**91**:3034-3040

[18] Zhou D, Wang R, Kamlah M. Determination of reversible and irreversible contributions to the polarization and strain response of soft PZT using the partial unloading method. Journal of the European Ceramic Society. 2010;**30**:2603-2615

[19] Zhou D, Kamlah M, Munz D. Effects of bias electric fields on the non-linear ferroelastic behavior of soft lead zirconate titanate piezoceramics. Journal of the American Ceramic Society. 2005;**88**:867-874

[20] Hwang S, Lynch C, McMeeking R. Ferroelectric/ferroelastic interactions and a polarization switching model. Acta Metallurgica et Materialia. 1995;**43**:2073-2084

[21] Kim K, Huber JE. In situ observation of ferroelastic domain evolution in a near-morphotropic $Pb(Zr,Ti)O_3$ ceramic by piezoresponse force microscopy. Journal of the European Ceramic Society. 2015;**35**:1459-1468

[22] Xu J, Xie S, Liu Y, Wang Q, Zhu J. Mechanical properties and multi-field ferroelastic response of Nb/Mn Co-doped $CaBi_4Ti_4O_{15}$ high-temperature ferroelectric ceramics. Ceramics International. 2024;**50**:50607-50621

[23] Moreno JC, Guiu F, Meredith M, Reece M, Alford NM, Penn S. Anisotropic and cyclic mechanical properties of piezoelectrics-compression testing. Journal of the European Ceramic Society. 1999;**19**:1321-1324

[24] Calderon-Moreno JM. Stress induced domain switching of PZT in compression tests. Materials Science and Engineering: A. 2001;**315**:227-230

[25] Okayasu M, Ozeki G, Mizuno M. Fatigue failure characteristics of lead zirconate titanate piezoelectric ceramics. Journal of the European Ceramic Society. 2010;**30**:713-725

[26] Marsilius M, Granzow T, Jones JL. Quantitative comparison between the degree of domain orientation and nonlinear properties of a PZT ceramic during electrical and mechanical loading. Journal of Materials Research. 2011;**26**:1126-1132

[27] Okayasu M, Bamba K. Domain switching characteristics of lead zirconate titanate piezoelectric ceramics on a nanoscopic scale. Journal of the European Ceramic Society. 2017;**37**:145-159

[28] Gao P, Britson J, Nelson CT, Jokisaari JR, Duan C, Trassin M, et al. Ferroelastic domain switching dynamics under electrical and mechanical excitations. Nature Communications. 2014;**5**:3801

[29] Forrester J, Kisi E, Studer A. Direct observation of ferroelastic domain switching in polycrystalline $BaTiO_3$

using in situ neutron diffraction. Journal of the European Ceramic Society. 2005;**25**:447-454

[30] Munoz-Saldana J, Schneider G, Eng L. Stress induced movement of ferroelastic domain walls in $BaTiO_3$ single crystals evaluated by scanning force microscopy. Surface Science. 2001;**480**:L402-LL10

[31] Tan Z, Xie S, Jiang L, Xing J, Chen Y, Zhu J, et al. Oxygen octahedron tilting, electrical properties and mechanical behaviors in alkali niobate-based lead-free piezoelectric ceramics. Journal of Materiomics. 2019;**5**:372-384

[32] Chen Y, Miao C, Xie S, Xu L, Wang Q, Zhu J, et al. Fracture behaviors and ferroelastic deformation in W/Cr Co-doped $Bi_4Ti_3O_{12}$ ceramics. Journal of the American Ceramic Society. 2016;**99**:2103-2109

[33] Chen Y, Miao C, Xie S, Xu L, Wang Q, Zhu J, et al. Microstructural evolutions, elastic properties and mechanical behaviors of W/Cr Co-doped $Bi_4Ti_3O_{12}$ ceramics. Materials & Design. 2016;**90**:628-634

[34] Chen Y, Xu J, Xie S, Nie R, Yuan J, Wang Q, et al. Failure mode, ferroelastic behavior and toughening effect of bismuth titanate ferroelectric ceramics under uniaxial compression load. Materials & Design. 2018;**152**:54-64

[35] Chen S, Wang T, Xie S, Leng J, Zhu Q, Li K, et al. High-entropy strategy for improved mechanical and energy storage properties in $BaTiO_3$–$BiFeO_3$-based ceramics. ACS Applied Materials & Interfaces. 2024;**16**:12521-12533

[36] Chen S, Wang T, Wang X-L, Li K, Zhu Q-F, Gong W-P, et al. Structural origin of enhanced storage energy performance and robust mechanical property in A-site disordered high-entropy ceramics. Rare Metals. 2025;**44**(1):551-564

[37] Martin A, Khansur NH, Webber KG. Electric field-induced changes in the ferroelastic behavior of $(Na_{1/2}Bi_{1/2})TiO_3$-$BaTiO_3$. Journal of the European Ceramic Society. 2018;**38**:4623-4630

[38] Shi X, Eckstein U, Lang S, Cicconi MR, Khansur NH. Temperature-dependent ferroelastic behaviour of antiferroelectric $AgNbO_3$. Acta Materialia. 2022;**232**:117931

[39] Strelcov E, Dong Q, Li T, Chae J, Shao Y, Deng Y, et al. $CH_3NH_3PbI_3$ perovskites: Ferroelasticity revealed. Science Advances. 2017;**3**:e1602165

[40] Li Y, Chu K, Liu C, Jiang P, Qu K, Gao P, et al. Superelastic oxide micropillars enabled by surface tension–modulated 90 domain switching with excellent fatigue resistance. Proceedings of the National Academy of Sciences. 2021;**118**:e2025255118

[41] Chen Z, Huang Q, Wang F, Ringer SP, Luo H, Liao X. Stress-induced reversible and irreversible ferroelectric domain switching. Applied Physics Letters. 2018;**112**:152901

[42] Bermejo R, Deluca M. Mechanical characterization of PZT ceramics for multilayer piezoelectric actuators. Journal of Ceramic Science and Technology. 2012;**3**:159-168

[43] Denkhaus SM, Vögler M, Novak N, Rödel J. Short crack fracture toughness in $(1-x)(Na_{1/2}Bi_{1/2})TiO_3$–$xBaTiO_3$ relaxor ferroelectrics. Journal of the American Ceramic Society. 2017;**100**:4760-4769

[44] Fang F, Yang W. Poling-enhanced fracture resistance of lead zirconate

titanate ferroelectric ceramics. Materials Letters. 2000;**46**:131-135

[45] Chen Y, Xu J, Xu Q, Xie S, Wang Q, Zhu J. Ferroelastic domain switching and R-curve behavior in lead zirconate titanate (Zr/Ti= 52/48)-based ferroelectric ceramics. Journal of the American Ceramic Society. 2020;**103**:1067-1078

[46] Seo Y-H, Vögler M, Isaia D, Aulbach E, Rödel J, Webber KG. Temperature-dependent R-curve behavior of $Pb(Zr_{1-x}Ti_x)O_3$. Acta Materialia. 2013;**61**:6418-6427

Chapter 2

Domain and Domain Wall Motion

Shedrach Yakubu

Abstract

Generally, domains in ferroic materials are a special kind of twin defect that occurs to counter the effect of the stray fields generated during phase transformation from a highly symmetric phase to a low symmetric phase. Domain walls serve as a boundary between two or more domains, either antiparallel (forming 180° domain wall) or perpendicular (forming 90° domain wall), based on the direction of the order parameter. The science and engineering of domain and domain wall motions are highly significant as this plays a crucial role in hysteresis modification and property enhancement in ferroic materials. This chapter shall attempt to capture some of the fundamental concepts of domain and domain wall formation and the types/kinds of domain walls discussed based on the nature of the change of the polar order at the domain wall plane. Domain wall motions, factors that restrict this motion, and the contribution of domain wall motion to switching polarization vectors will also be addressed. Various domain engineering techniques in ferroelectric single crystals, ceramics, polymers, and composites, as well as moods of applications such as memory storage, capacitors, and sensors, shall be explored in this chapter.

Keywords: domain, ferroelectricity, depolarizing field, polarization, domain engineering

1. Introduction

The science and engineering of domain and domain wall motion have been carefully explored recently by researchers [1, 2], especially regarding the contribution to polarization reversal in the ferroic class of materials. Most of the functional applications of ferroic materials, especially ceramics, thin films, and single crystals in telecommunication, energy storage, memory, sensing, transducing, and actuating devices, strongly depend on ferromagnetic, ferroelectric, and ferroelastic switching of magnetic spin, polarization vector, and elastic strain respectively, which are embedded functions of domain and domain wall motion [3–5]. Domains have long been considered a special kind of twin defect that occurs during phase transformation from a high symmetric phase to a low symmetric one, otherwise referred to as transformation twin [6], which is common in materials with a ferroic status. During this phase transformation, the low symmetry phase is at liberty to assume several states either by ionic displacement as in ferroelectric, shape change by structural deformation as in ferroelastic, or change in a magnetic moment as in ferromagnetic giving rise to homogeneous regions of domains [7]. Phase transition is a significant aspect of materials that plays a crucial role in property adjustment through structural modification; therefore, it is crucial for full understanding of the genesis of domain and domain wall formation.

Several concepts have been proposed to understand why and how domain and domain walls are formed in ferroelectrics. One of these is the theory of depolarizing field introduced by Evans et al. [8]. This theory describes domain wall formation by considering the influence of stray fields formed by surface bond charges perpendicular to the polarization vector destabilizing the energy of the ferroelectric material during phase transformation. In this sense, the effect of the stray field is minimized when the ferroelectric material breaks the polarized area into various homogeneous regions called domains in which the polarization vector points toward a specified direction. The boundary or interface between these domains is called domain walls, which have been suggested to be chemically rich and of the order of >10 nm in thickness [9]. In ferromagnetic materials, the driving force for the formation of a domain is an intrinsic attempt by the material to minimize the magnetostatic energy to attain stability in terms of energy contributions.

The physical response of ferroic material to external stimuli such as optical load, mechanical stress, and/or electric field is directly related to the domain pattern [10]. When such external stimuli exceed a critical value, it influences the vibration energy of the domain wall triggering domain wall motion. The hysteresis loop peculiar to these materials is clear evidence of domain wall motion [11], and the contribution to ferroelectricity cannot be overemphasized. According to Schultheiß et al. [12], in an applied electric field, the domain wall motion proceeds in three main steps, which include the nucleation process at the surface of the ferroic material followed by the growth process, which may be deep down (forward) or sideways, and finally, coalescence of various domains to give rise to a single domain. The domain and domain wall can be engineered to enhance certain physical properties directly related to ferroelectric, ferromagnetic, and ferroelastic hysteresis; this is achievable via the understanding of the domain configuration in terms of the orientation of the polar order, size of the domain, and the domain structure [13]. Impurity doping (donor or acceptor doping) of Pb-based ABO_3 perovskite ferroelectrics like PMN-PT can engineer the domain by generating either O^{2-}-vacancies or Pb^{2+}-vacancies thus influencing high-power applications. Donor or Acceptor doping of perovskites engineered the domain walls to obtain soft or hard perovskites with a reduced or enhanced coercive field, respectively. The domain wall can be engineered by considering the charge conduction, mobility, and kinetics of the domain wall [13]. Studies have shown that the domain wall motion is the frictional generating thermal energy that is dissipated and thus influences the ferroic behavior [13].

The ferroic class includes ferromagnetic, ferroelastic, and ferroelectric order; therefore, for convenience, the analogy and illustration of domain and domain wall motion in this chapter shall be explored mostly in terms of ferroelectricity with little attention to ferromagnetic and ferroelastic domain walls. This will be helpful for readers in that ferroelectric perovskites are very common with significant promising properties domiciled in the domain structures, thus attracting a lot of attention. This will also make it easy to understand certain domain wall concepts related to symmetrical breakdowns during phase transformations, spontaneous polarization, and switching of polarization vectors. The phenomenon of domain wall in ferroelectrics is like ferromagnetic and ferroelastic with subtle differences. The reader is employed to pay attention to the paraelectric to ferroelectric, paramagnetic to ferromagnetic, and paraelastic to ferroelastic phase transition in ferroelectrics, ferromagnetic, and ferroelastic materials, respectively, as this plays a crucial role in understanding domain formation which occurs to stabilize the ferroic material from the influence of strain, electrostatic and magnetostatic energy contributions generated during these transitions.

1.1 The ferroic class from crystallographic viewpoints

The ferroic class as stated, is a class of material that responds to an applied field by the reversal of spontaneous polarization as in ferroelectrics, spontaneous magnetization as is the case with ferromagnetics, and spontaneous strain as is the case with ferroelastics. Ferroelectrics are influenced by structural symmetry, which significantly influences their properties, including dielectric, piezoelectric, and other non-linear ferroic behaviors [14]. Except for translational symmetry, there are 32 crystal point groups, out of which 11 are centrosymmetric and 21 non-centrosymmetric. 20 out of the 21 exhibit a piezoelectric effect, varying with direction due to material anisotropy, especially in single crystals [15]. 10 out of the 20 piezoelectric crystals are isotropic and, when polarized by stimulus in a specific crystallographic direction, result in a piezoelectric effect. Still, when they are subjected to thermal energy, the overall piezoelectric effect cancels out since they are isotropic, and thermal energy is equivalent in all crystallographic directions [16]. The other 10 are anisotropic crystals, such that applied thermal energy which is equivalent in all crystallographic directions, only favors a single crystalographic direction, causing polarization in the bulk of the crystal and thus called pyroelectric effects. Pyroelectric crystals spontaneously polarize at room temperature [17]. When the polarization of a pyroelectric crystal is reversible by an applied electric field, it becomes a ferroelectric crystal [18]. The same is true for ferroelastic, as every polarization reversal retains a strain in the material, giving rise to ferroelastic effects. For a material to be ferroelectric, it must undergo spontaneous polarization, which must be reversible in an applied electric field. The criteria for this is the existence of non-centrosymmetry in the crystal structure.

1.2 Spontaneous polarization in ferroelectric crystals

Polarization in a ferroelectric crystal is dominated by ionic displacement from one lattice position to the other, resulting in the separation of the centers of cations and anions due to lattice vibration [14]. At room temperature, ion vibration in ferroelectric crystals causes charge separation and, hence, spontaneous polarization. Understanding this phenomenon requires considering factors that maintain polarization and crystal stability. Every ion in ferroelectric crystal possesses a local electric field E^{loc}, whose strength is proportional to the extent of polarization P when displaced by lattice vibration: If there are N numbers of ions displaced by the effect of lattice vibration, causing polarization at room temperature, the total Gibb's dipole energy contribution resulting from the product of the local electric field and induced dipole moment for the N ions is given thus:

$$G_{dip} = -N\left(\frac{\alpha\gamma^2}{9\varepsilon_o^2}\right)P^2 \quad (1)$$

where γ is the Lorentz factor and for an isotropic cubic system γ is equal to 1, α is the polarizability tensor, ε_0 the permittivity of free space. The ions occupying lattice positions are said to be vibrating about their mean position. If this lattice vibration results in a displacement x, then there must be an elastic energy contribution G_{ela} related to the ionic displacement carrying charge q, and thus expressed in terms of the kinetic energy of the ions and temperature-dependent force constants K and $K^{\wedge}$:

$$G_{ela} = \left[\left(\frac{K}{2q^2N} \right) P^2 + \left(\frac{K^{\wedge}}{4q^4N^3} \right) P^4 \right] \quad (2)$$

The total free Gibbs energy contribution in this case will be the summation G_{dip} and G_{ela}, which define thermodynamically the stability state of the crystal when polarized:

$$G_{tot} = \left[\left(\frac{K}{2q^2N} \right) P^2 - N \left(\frac{\alpha\gamma^2}{9\varepsilon_o^2} \right) P^2 + \left(\frac{K^{\wedge}}{4q^4N^3} \right) P^4 \right] \quad (3)$$

The contribution of $K^{\wedge}$ is thermodynamically significant only at very high temperatures. When the coefficient of the elastic energy contribution is equal to or greater than the coefficient of the dipole energy contribution, the polarization becomes zero, and the paraelectric phase is realized, whereas when the coefficient of the dipole energy contribution is higher, then the polarization becomes.

$$P^2 = \left[\frac{N \left(\frac{\alpha\gamma^2}{9\varepsilon_o^2} \right) - \left(\frac{K}{2q^2N} \right)}{\left(\frac{K^{\wedge}}{4q^4N^3} \right)} \right] \quad (4)$$

If this condition is true at room temperature, the crystal becomes spontaneously polarized. The Lorentz factor significantly influences spontaneous polarization in perovskite crystals like PZT-PT, PMN-PT, and ($BaTiO_3$ has $\gamma = 10^2$ [14]), enhancing the dipole energy coefficient and causing a high magnitude of polarization, indicating ferroelectric properties. For a cubic system, the Lorentz factor is almost one; hence, it has a paraelectric nature [14]. The generation of the spontaneous polarization in a ferroic material can be such that the polarization vector aligns in several directions at different angles, giving rise to domains and domain walls.

1.2.1 Antiferroelectricity

The case of ferroelectricity is considered a spontaneous lattice polarization in that the overall bulk polarization is non-zero, and the reversal gives a single loop hysteresis [11, 19]. In the antiferroelectric crystal, the direction of the dipole moment in each sublattice points in the opposite direction to that of an adjacent sublattice, resulting in the cancelation of the overall bulk polarization [20]. In essence, if we consider two sublattices with polarization P_a and P_b, then the sublattice polarization in the ferroelectric phase can be related as; $P_a = P_b$ whereas that of the antiferroelectric phase is given as; $P_a = -P_b$.

The free energy difference between the ferroelectric phase and the antiferroelectric phase is very small such that at a particular applied electric field or mechanical

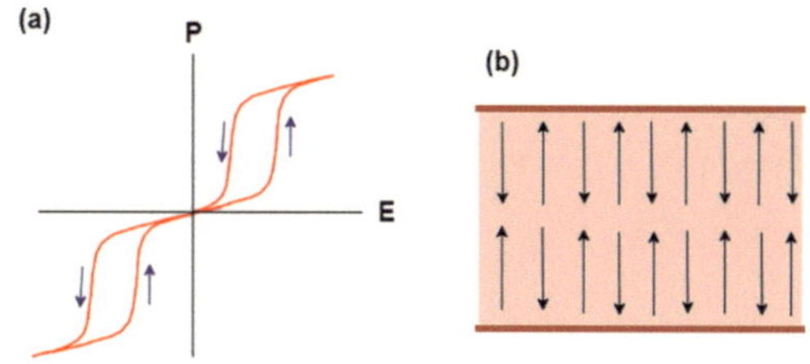

Figure 1.
Double hysteresis (a) and antiparallel sublattice dipole (b) in antiferroelectric.

stress, antiferroelectric-ferroelectric phase transition becomes possible, and the evidence is the associated double loop hysteresis shown in **Figure 1**. Other evidence of antiferroelectric-ferroelectric phase transition is a giant volume change and a high increase in dielectric polarization.

2. Domain wall formation in ferroic materials

The idea of domain wall formation borders on phase transitions either from paraelectric to ferroelectric or between two ferroelectric stats as in ferroelectric materials, or from paramagnetic to ferromagnetic or between two ferromagnetic phases as in ferromagnetic materials or from one ferroelastic state to another as in ferroelastic materials. This kind of phase transition significantly breaks the symmetry of the material. To compensate for the associated energy contributions that may hamper the stability of the crystal, the bulk of the ferroic material breaks into homogeneous regions with lower energy contributions, as we shall see.

2.1 Ferroelectric domain wall formation

The concept of domain and domain wall formation in the ferroelectric crystal is directly related to the depolarizing field. The surfaces of most ferroelectric crystals perpendicular to the direction of ferroelectric polarization contain bond charges that generate a strong depolarizing field d_p, that acts in the opposite direction to that of the ferroelectric polarization P [8]. The energy contribution of the depolarizing field (electrostatic energy) tends to perturb the stability of the ferroelectric crystal and therefore, to minimize this energy, the crystal breaks into various regions of homogeneous polarity pointing at a unique direction [6]. Each region of homogeneous polarity is called a domain and the boundary between them is regarded as a domain wall. This idea is illustrated in **Figure 1**, where a monodomain crystal is shown with a very high depolarizing field (**Figure 2a**), when the monodomain breaks down into two (**Figure 2b**), the depolarizing field reduces tremendously and is thus eliminated in a multiple domain (**Figure 2c**).

2.2 Ferromagnetic domain wall formation

In ferromagnetic materials, a domain wall can be formed in the same manner as in the case with ferroelectric, but in this case, the alternating field to the magnetized field is the demagnetizing field, which impairs the stability of the crystal. This also influences

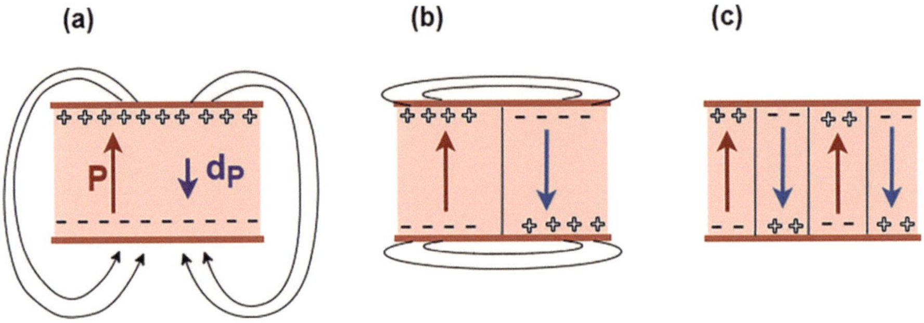

Figure 2.
Formation of domain and domain wall: (a) strong depolarizing field on ferroelectric with a single domain, (b) breakdown of the single domain into 180° domain wall reduces the depolarization field, and (c) formation of multiple 180° domain walls eliminates the depolarization field.

the ferromagnetic material to break its single domain into several domains in other to eliminate the influence of the demagnetizing field. The energy related to the demagnetizing field is magnetostatic energy, which must be minimized through domain formation to stabilize the bulk ferromagnet. The sum of the domain's magnetization is often zero, which shows that when several domains are formed in a ferromagnet, the material loses its magnetization outside an applied magnetic field. This process can be illustrated as in ferroelectric (**Figure 2**); however, the associated field is magnetic.

2.3 Ferroelastic domain wall formation

Inherent energy related to ferroelectric distortion can also cause the formation of domain and domain walls, this time, shear directions are of interest [21]. For example, in terms of strain related to a phase transition, perhaps from paraelectric to ferroelectric or from one ferroelectric to another ferroelectric phase, the internal strain during the shear effect may not be distributed evenly within the material and to minimize the energy contribution related to the strain, the shear direction may be alternated giving rise to a twin-like boundary with each region of like shear forming ellipsoidal elastic dipoles referred to as elastic domain [21]. This kind of domain formation is dominant in materials that also undergo ferroelastic phase transition, where there is a change in one of the tensorial thermal expansion coefficients related to the crystal structure and shape of the material, which retains strain during the high-low symmetry transition. Retained strains are often in the form of ellipsoidal elastic dipoles, as shown in **Figure 3**.

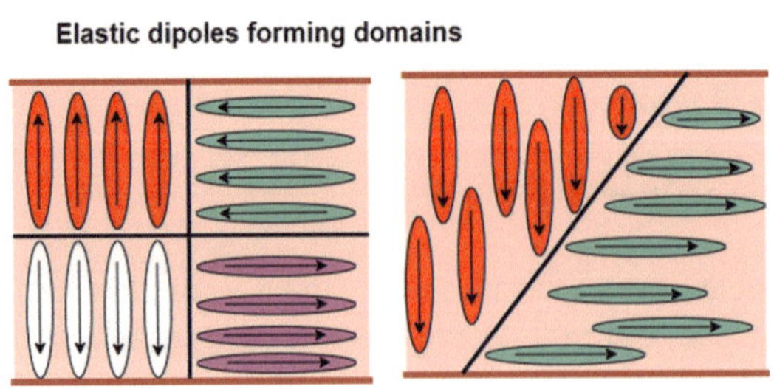

Figure 3.
Ellipsoidal dipoles forming elastic domain and domain wall.

2.4 Domain wall sizes and implications

According to Kittel's law [20], the width of the domain in terms of its size is directly proportional to the square root of the inverse of its thickness. The domain size in general depends on the competing energy difference between the domain and domain wall. The energy of the domain depends on its width, and that of the domain wall depends on its number. To minimize this energy contribution, the domain size w, most follow the Kittel's law stated mathematically as:

$$\mathrm{w} = \sqrt{\frac{\sigma}{\mathrm{U\,d}}} \tag{5}$$

where σ is the energy of the domain wall, U is the volumetric energy, and d is the thickness of the bulk material. In ferromagnetic materials, magnetic energy exchange promotes a wide domain wall, and ferromagnetic anisotropy promotes a finer domain wall, thus influencing the shape and size of the ferromagnetic domain and domain wall. In ferroelectrics, dipole-dipole energy and ferroelectric anisotropy are of the same order of magnitude, and the domain walls are finer in ferroelectrics than in ferromagnetic ones. The fine domain structures, in terms of their sizes, significantly influence the physical properties of the ferroic material.

2.5 Domain wall visualization

Research today concerning domain and domain walls in ferroelectric and ferromagnetic single crystals is made possible due to the possibility of visualizing the domain and domain wall in an in-situ experiment with spatial resolution up to subatomic levels. A good number of techniques have been developed, all of which stem from microscopy, ranging from optical, electron, and scanning probe microscopy. Optical visualization of the domain is possible either by birefringence contrast and chemical etching in ferroelectrics or by magneto-optical Kerr effect in ferromagnetics. Interactions of electrons with the surfaces of ferroic materials can also generate information through backscatter or secondary electrons to provide images of the various domains present. Some of these electron microscopes include scanning electron microscope (SEM), transfer electron microscope (TEM), and scanning transfer electron microscope (STEM). In some cases, surface functionalization of the ferroic sample is also possible in an electron microscope for domain wall visualization. The most effective technique for the visualization of the domain walls is through scanning probe microscopy, which includes atomic force microscopy (AFM) that takes advantage of Van-der-wall forces at the surface, and scanning tunneling current microscopy (STCM) that embodies electrostatic force microscopy (EFM), magnetic force microscopy (MFM) which rely on influences from the magnetic field, piezoelectric force microscopy (PFM) which relies on non-linear polarization response to an applied electric field, and photoemission electron microscopy (PEEM) which takes advantage of the asymmetric nature of the electron charge distributions within the surfaces of the ferroic material.

The above techniques have been successfully used to study materials with a single ferroic order. However, significant interference occurs when two or three ferroic orders are coupled in a material. For instance, consider an MFM investigation of a domain in a multiferroic with coupled ferromagnetic and ferroelectric potentials. It becomes difficult to probe the magnetic field without interference from the polarization field and vice vasa. This limitation makes it difficult to visualize domain and domain walls in multiferroic materials using scanning probe microscopy. Hence, techniques like second harmonic generation (SHG) that rely on doubling of light frequency in sample materials, makes it suitable for visualization of independent magnetic and electrical fields in multiferroics. Other researchers employ the combination of two or more probing techniques to come to a suitable conclusion regarding domain structures in multiferroics.

3. Domain wall geometry

Domains contain polarization vectors and these vectors in adjacent domains can either align in an antiparallel manner creating a domain wall of 180° (**Figure 4a**) or in a nearly perpendicular manner creating a domain wall of 90° (**Figure 4b**). In certain ferroelectric materials like tetragonal $BaTiO_3$, 71° and 109° domain walls can be observed; however, for convenience, they are all categorized as 90° domain walls [21]. In 90° domain walls, adjacent domains' polarization or magnetization vectors can be coupled via head-to-tail, tail-to-tail, or head-to-head. The most common one found in perovskites like $BaTiO_3$ is head-to-tail, as shown in **Figure 2b**. This kind of arrangement influences the nature of charges found within the vicinity of the domain wall, making the physical properties within the wall different from that of the domain itself. Tail-to-tail and head-to-head configurations often result from the interactions between 180° and 90° walls, forming a new 90° wall. These configurations of 90° domain walls show no significant differences in terms of optical properties for visualization; however, they can be differentiated using a multiple interferometric technique [22]. In some cases, a polarized microscope can provide information about 90° domains due to differences in spontaneous birefringence. It should be noted that this technique is not efficient for visualizing 180° domain walls where antiparallel polarization vectors show similar optical responses. To visualize an 180° domain wall, an electric field and strain must be induced to cause some infinitesimal polarization reversal in two domains with opposing directions, thus providing different spontaneous birefringence. It has been shown that geometrically, both 180° and 90° domain walls can either be a straight line or cylindrical, forming a curved ring-like domain wall. These curvatures greatly impact the properties of the domain wall, making them significantly different from the domain, as we shall see.

3.1 Implication of the domain wall geometry

In recent discoveries, the domain wall including 180° and 90° walls can either be straight or curved. The curvature of the domain wall has a significant influence on the physical properties of the domain. It must be first established that the properties of

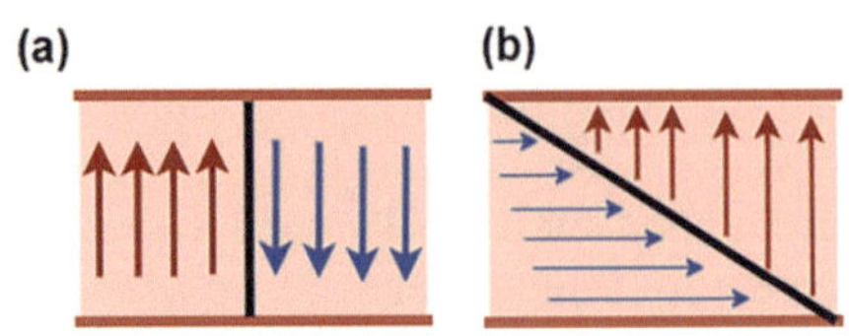

Figure 4.
180° (a) and 90° (b) domain walls.

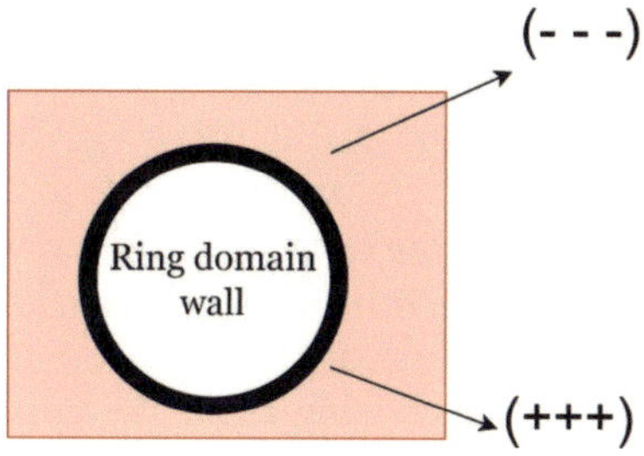

Figure 5.
Ring domain wall.

the domain are significantly different from those of the domain walls in that some of the walls possess electrical conductivity, which is a property alien to the domain itself since the ferroic material is considered a poor conductor of electricity. Researchers have also demonstrated intrinsic hysteresis within the domain wall, memory effects, and time-dependent dynamics [23]. This observation (domain wall charge conduction) implicates the domain dynamics as a function of polarization dynamics in terms of shape, geometrical change, or ionic processes involving the vacancy/carrier separation process. In a clear term, the electrical conductivity of the domains is greatly influenced by the geometry of the domain, which is either straight or curved. For straight domain walls, the electrical conductivity does not vary at any point, and this has been studied by Vasudevan et al. [23], whereas curved domain walls, often in the form of a ring wall (**Figure 5**), show various electrical conduction dynamics at different curvature positions. The electrical conduction dynamics due to domain wall curvature also influence the polarization dynamics differently; hence, the ferroelectric properties are affected. An overview of a charged domain wall has been disclosed by Bednyakov et al. [24]. The discovery of electrical conduction on the domain wall leads to several investigations of the nature of the charge configurations on the domain wall and its electronic applications. The charge configurations of the domain wall have been explored for memory storage device fabrication [25].

4. Types of domain walls

There are three known kinds of domain walls, which are classified based on the nature of changes in the polar order at the domain wall plane which exist in both ferroelectric and ferromagnetic materials [8]. These domain walls include Ising-type domain walls, Bloch-type domain walls, and Neel-type domain walls.

4.1 Ising-type domain wall

In an Ising-type wall, the polar axis is fixed and moving across the wall in such a manner that changes the magnitude of the polarization as it moves across the plane of the domain wall from up to down and passes through zero at the center. These walls are the finest and are found predominantly in ferroelectric materials. Geometrically, 180° domain walls in ferroelectrics are assumed to be majorly of the Ising-type because the coupling for instance between polarization and strain in the ferroelectrics requires a large amount of energy for the polarization vector to rotate away from the plane of the symmetry allowed directions hence an Ising-type domain is preferred as for 180° domain wall in ferroelectrics. However, it is possible to find a mixed situation where other types of domain walls like the Bloch-type co-exist with the Ising-type in 180° domain walls. This often occurs during phase transitions on cooling where the Ising-type domain in ferroelectric perovskites tends to switch by diminishing polarization magnitude and at the same time allowing a switch that rotates across the plane of the domain wall, resulting in mixed Ising-Bloch domain walls [26]. Lee et al. [27] theoretically, using the density function, predicted both PbTiO3 and LiNbO3 perovskite materials to show the mixed Ising-Bloch domain character. It is important to know that such a mixed domain wall can exhibit physical properties dissimilar to those with a single-type domain wall.

4.2 Bloch-type domain wall

In a Bloch-type wall, the polarization rotates within the plane of the domain wall without changing its magnitude. In this type of domain wall, charges are domiciled on the external surfaces and interface between the two domains and often occur in bulk materials with uniaxially anisotropic plane axes. The idea of the Bloch-type domain wall is that the transformation or polarization switch within the wall is continuous and not abrupt, with several magnetic spin or polarization vectors lying orthogonal to each other such that the exchange energy required for stability can be lowered when

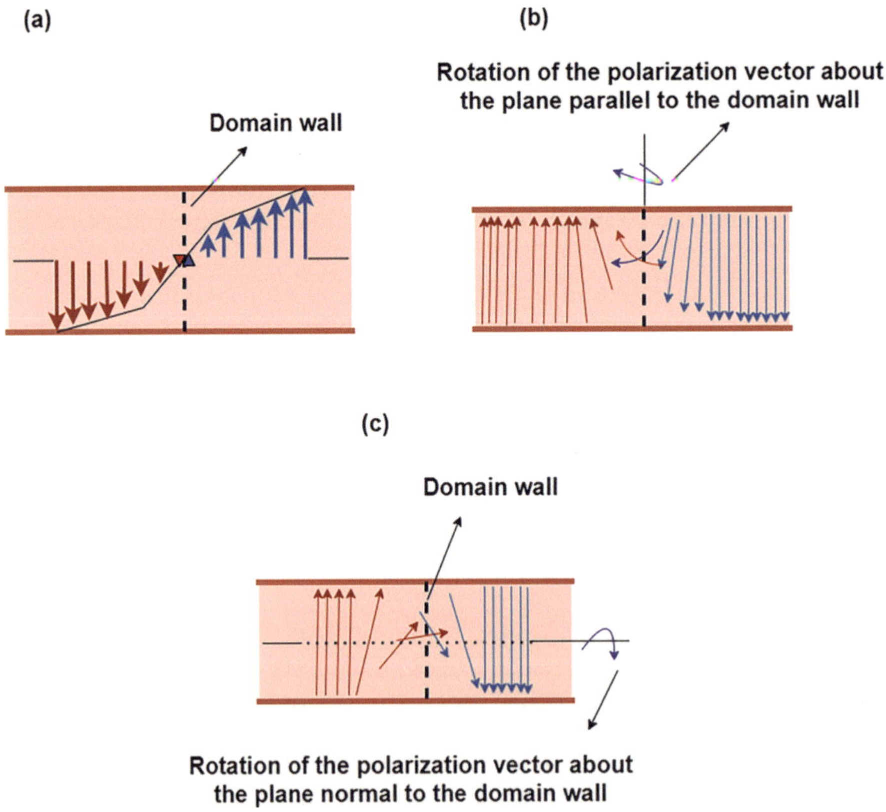

Figure 6.
Domain walls based on the change in polar directions. (a) Ising-type domain wall, (b) Bloch-type domain wall, and (c) Neel-type domain wall.

distributed among several polarization vectors or magnetic spin within the wall. The Bloch-type domain exists in ferroelectrics subjected to high stress to overcome the effect of the electrostrictive energy, allowing rotation in the disallowed symmetry directions and naturally in ferromagnetic materials, making it crucial for domain studies and applications. In $BaTiO_3$, the Bloch-type domain exists in the 4 mm space group of the tetragonal crystal system predicted by Monte Carlo calculations [28]. They observed a distortion in the orthogonal plane of the domain wall; however, they considered it a computational error.

4.3 Neel-type domain wall

In the Neel-type wall, the polarization is rotated perpendicular or normal to the polar axis, and at the center, they lie in the orthogonal direction to the plane of the domain wall. Both Bloch and Neel-type domain walls are predominantly found in ferromagnetic materials. The Neel domain walls present special domains that depend on the thickness of the ferromagnet. A Transverse Domain (TD) wall is possible for thin ferromagnets. This domain operates so that the spin moment is rotated within the plane of the sample thickness, thus minimizing the competition between the energy contributions of the demagnetizing field and that of the magnetization field, enhancing stability in the bulk material. The domain can be transformed into a Vortex Domain (VD) by increasing the thickness of the ferromagnetic material. In the VD, the magnetic spin forms spiral-like structures that attempt to minimize the imposed energy contribution of the demagnetizing field [29]. **Figure 6** illustrates the three domain wall types in ferroelectric and ferromagnetic materials.

5. Domain wall motion

The major evidence that a domain and domain wall are formed in a ferroic material is its contribution to ferroelectric hysteresis that arises from domain wall motion. As outlined earlier, these motions can be visualized via techniques such as optical, scanning electron, scanning probe, second harmonic generation, and photoemission electron microscopy [8]. Generally, materials with the ferroic status undergo polarization reversal or domain switching when subjected to an applied field that accounts for the inherent hysteresis [11]. Domain wall motion plays a key role in polarization switching, facilitating the nucleation and growth of domains initially antiparallel to the applied field. Domain wall motion can be activated by an applied electric field, magnetic field, and stress field and thus proceeds by three well-known mechanisms including domain nucleation, domain growth either forward or sideward, and finally coalescence of adjacent domain walls [30].

Nucleation often occurs in a specific crystallographic direction where polarization is favored. In that direction, expansion of the domain occurs at the expense of other domains pointing to other crystallographic directions. For every switching process, nucleation will always arise in the favored direction, which specifies the inhomogeneity and anisotropic nature of the nucleation process, depending on the field in which it is applied. Domain growth occurs rapidly in the forward direction into the bulk of the material, followed by sideways growth. The process depends on electric, magnetic, and stress field strength, defects in material, and material geometry. In the final stage of domain wall motion, the coalescence of the domain occurs when two domains combine, forming a larger one and the wall disappearing between them.

Domain wall motion is always hampered by grain boundaries, especially in ceramic materials. Marincel et al. [31] studied domain wall motion across grain boundaries in thin films and discovered that the nature of the grain boundaries plays a crucial role for instance, a twisted grain boundary hampered the domain wall motion more when compared to a tilted grain boundary. From kinetics, the domain wall motion, as we shall see, is greatly influenced by defects present in the ferroic material. The most significant evidence of domain wall motion is the hysteresis loop in ferromagnetic and ferroelectric materials, whereas in ferroelastic materials, a butterfly loop becomes the signature. This hysteresis is presented in **Figure 7(a)** for polarization or magnetization and **Figure 7(b)** for mechanical strain.

5.1 Domain wall kinetics

The kinetics of the domain wall have been said to follow a Boltzmann's distribution where the velocity v at an applied field strength E, is of the form [12]:

$$v = v_o \exp^{-\left(\frac{E_a}{E}\right)} \tag{6}$$

where v_o the initial velocity in the absence of applied electric field and E_a is the activation energy. Materials with low mobile domain walls are considered hard, having very high coercive fields while those with fast traveling domain walls are regarded as soft, having lower coercive fields [21]. The kinds of doping (e.g., acceptor or donor doping) significantly influence the domain wall's mobility. Depending on the material application, doping can be used to engineer the domain wall motion. Soft ferroelectrics are usually suitable for switching applications, and since they have high dielectric and electromechanical constants, they can be used to convert mechanical energy to electric energy. Hard ferroelectrics on the other hand have high resistance to depoling and hence can be used for actuating functions in converting electrical energy to mechanical [21].

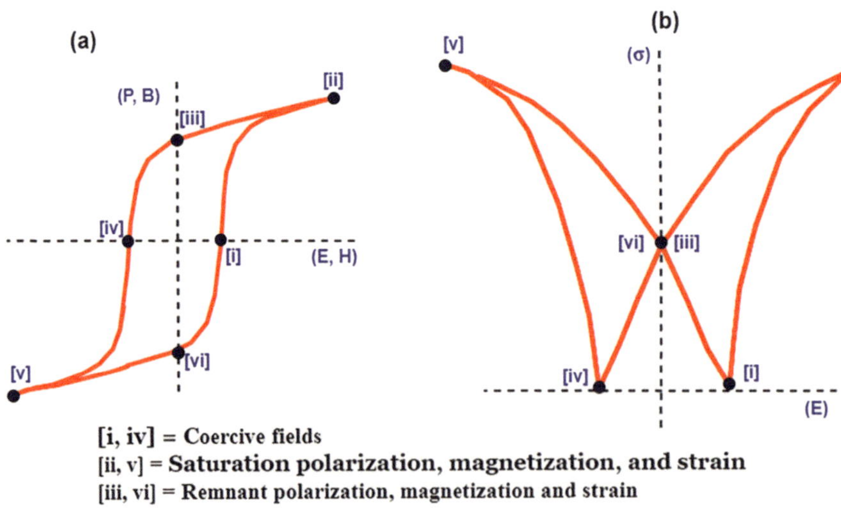

Figure 7.
Nature of hysteresis that arises during domain wall motion for ferroelectric/ferromagnetic hysteresis (a), and ferroelastic hysteresis (b).

6. Contribution of domain wall motion to ferroelectricity

When a ferroelectric crystal is subjected to a critical applied electric field called the coercive field, domain wall motion is first nucleated, thus growing forward and sidewise, followed by domain coalescence when the applied electric field exceeds the coercive field [12]. This process results in the poling of the most favored domain in the direction of the applied field, and the result is the hysteresis loop, where the applied electric field is first in the forward direction and then reversed. Four intrinsic and frequency-dependent contributors to polarization in ferroelectrics include electronic, atomic, ionic, and space charge. The fifth contributor, which is an extrinsic contributor to polarization in ferroelectric materials, is the domain wall motion [32]. Herbiet et al. [33] studied the volume contribution of the domain wall motions to ferroelectricity and thus discovered the piezoelectric and dielectric constants in ferroelectric materials are greatly influenced by the motion of non-180° domain walls. In the case of piezoelectrics, the material is first poled into a single domain, and then via the application of mechanical stress or electric field approximately 1/3 of the coercive field, the polarization shows a non-linear behavior with the applied field characterized by the piezoelectric coefficient.

If we consider a single crystal with various domain structures with zero applied fields, the domains in a ferroelectric material are pointed randomly in different directions to minimize the effect of the stray field (**Figure 8**). Point (a) presents a net zero polarization in the ferroelectric material due to the randomization of the domain polarization in the absence of an applied electric field whose resultant polarization is canceled out. When an electric field is applied to the material, a non-linear polarization response occurs until point (b), where the polarization becomes saturated. On switching the direction of the applied field, the domain will gradually reorient such that at zero applied field, there is still a remnant polarization

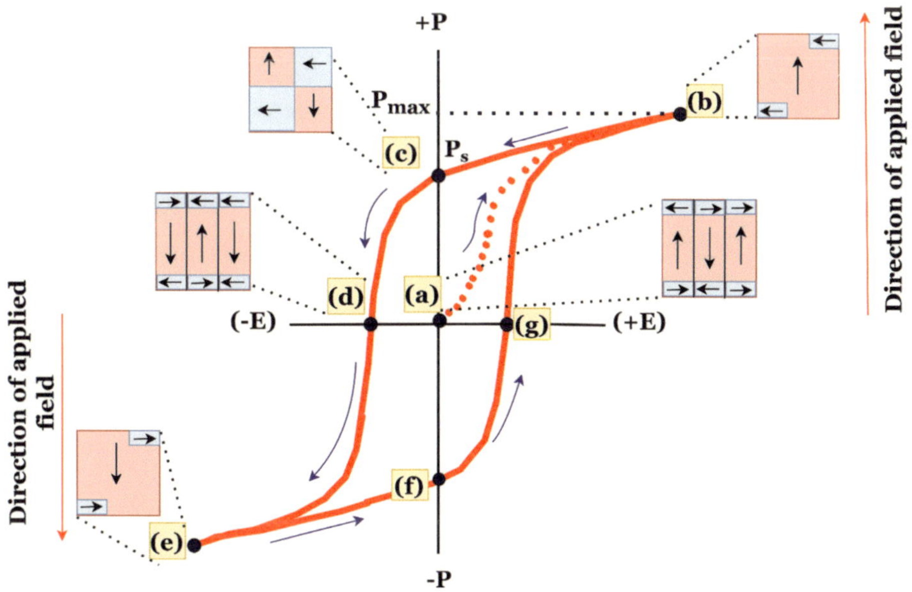

Figure 8.
Domain switching.

equivalent to point (c). Further, a decrease in the applied field until a point (d) called the coercive field in the reverse direction is reached where the polarization becomes zero. At saturation in the reverse direction, the polarization reaches its maximum, but this time, the polarization vector is pointed in the opposite direction to the initial vector, as shown in point (e). When the applied field is again reversed in the forward direction up to zero applied field, a remnant polarization in the negative axis is observed corresponding to the point (f). When a critical field called the coercive field is applied in the forward direction, reorientation of the domain or polarization vectors occurs. However, the polarization disappears and becomes zero due to the randomization of the domain wall, whose net polarization is zero, as shown in point (g).

From the hysteresis loop of **Figure 8**, the non-linear region between points 'a' and 'c' exists which corresponds to the non-linear response that depends on 90° domain wall motion and must often result in vibrational deformation of the bulk materials thus, characterizing the converse piezoelectric behavior. The reorientation of the polarization vector in this case is accompanied by domain wall motion until a saturation point where all domains have been successfully switched. At the saturation point, there is no increase in polarization with the applied field and a single domain is possible (points b and e). The hysteresis loop discussed in this perspective for ferroelectric is synanimouse to ferromagnetic, with the difference being in the fact that and applied magnetic field is used to induce the magnetization causing the hysteresis.

The effect of grain size on dielectric properties has been linked to domain wall motion in that, fine grain size, being perceived to have a single domain, has a low contribution to polarization compared to grains which, when above a particular critical size, are made up of multiple domains that contribute enormously to the polarization vector. At some larger grain sizes, the domain wall volume density decreases and therefore reduces the contribution to polarization [8]. The coercive field in the forward and backward direction is the same and therefore, entails the frictional nature of the domain wall motion [21], which gives rise to a strain in a particular crystallographic direction when poled, as shown in **Figure 7**. An alteration in terms of applied stress or electric field results in direct or converse piezoelectricity, respectively [21]. Research has shown that not all domains can be switched in a particular direction, in fact, as presented in **Figure 9**, only domains that lie within the cone of acceptability in the direction of the applied electric field can be poled. The region of the forbidden domain switch remains frozen until the point of dielectric breakdown of the ferroelectric [34].

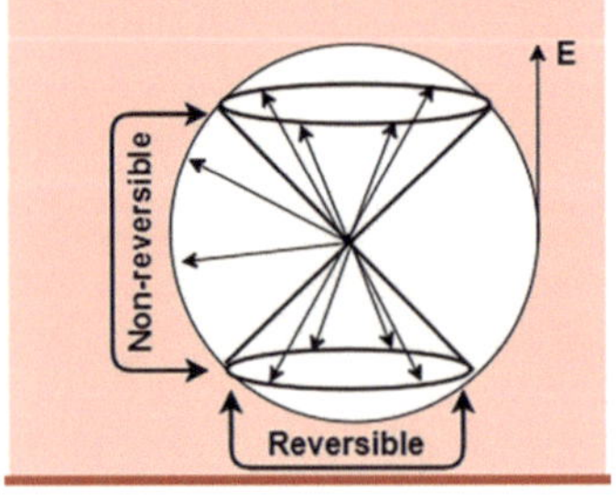

Figure 9.
Region of polarization reversal in ferroelectrics.

DOI: http://dx.doi.org/10.5772/intechopen.1006892

6.1 Factors affecting domain and domain wall motion

Ferroelectric domains are a function of polarization, which strongly depends on the crystal symmetry of the material under certain thermodynamic conditions. Generally, these conditions spang through temperatures, pressures, and applied electric fields, which induced noncentral-symmetry in the material either by lattice distortion, ionic displacement, and so on, capable of generating dipole moments and thus polarization. In some cases, when these dipoles are generated, they are subject to the influence of local fields, chemical doping, dislocations, and grain effects. Some of these factors are considered briefly. The width of the domain wall can also be influenced by the local field, which either increases the energy of the potential well or increases it.

6.2 Local field

Local fields generally exist in ferroic materials and greatly influence parameters like the domain wall mobility, kinetics, width, energy, structure, and even pinning and depinning of the domain walls. The domain wall structure, especially when tilted or curved can interact with the local fields, which often provide significant information in understanding the domain wall dynamics which is characterized by domain nucleation, growth or shrinkage, and vibration of domain walls under applied fields. Apart from the tendencies of pinning and depinning of domain walls by local field, it is also possible to initiate domain wall motion under this field causing a slight domain shift via the motion or migration of ionic species under applied fields causing separation and polarization as shown in **Figure 10** [35].

6.3 Lattice distortion

Lattice distortion plays a significant role in domain and domain walls, especially in terms of domain formation, domain width, mobility, and kinetics. Distortions in crystal lattice often arise from the influence of either a donor or acceptor dopant, retained strain from either thermal or stress fields, and so on. Within the domain walls, the variations in the order parameters have been proven via first-principles calculations to be due to large lattice distortion within the vicinity of the domain walls [36]. The gradual

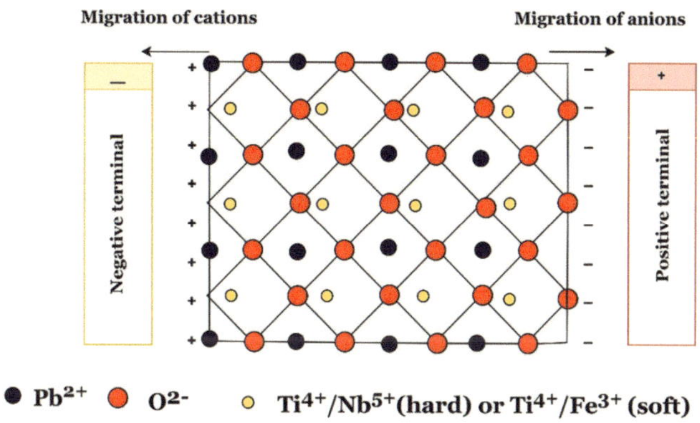

Figure 10.
Effects of applied fields on ferroelectrics.

polarization switching zone observed in ferroelectric material with wide domain width is caused by lattice distortion imposed on the domain wall, resulting in domain width widening. This is highly significant as it prevents the abrupt nature of polarization reversal and rather spreads the switch within the wide region of the domain width, creating a well-defined hysteresis in ferroelectric single crystals. Distorted lattice in domain walls often serves as pinning centers for domain wall motions, especially in a region with highly dense lattice distortion, as shown in **Figure 11(a)**. These effects can greatly influence the dynamics and kinetics of mobility of the domain wall, as enshrined in Shi et al. [37]. It is also possible for the impurity or dopants present in the material to interact with the fields of the lattice distorted region, thereby modifying the local electronic fields of the material and thus the domain energy landscape [35].

6.4 Crystal structure

The crystal structure of most multiferroic materials plays a significant role in modifying the order parameter as well as the nature of the domain and domain wall configuration. Generally, the high-temperature cubic systems are paraelectric and show no sign of spontaneous polarization, as shown in the cubic undistorted cell in **Figure 11(b)** [38]. However, on the transition to lower symmetry crystal systems like the tetragonal, monoclinic, orthorhombic, and rhombohedral, a non-centrosymmetry set in, and this results in the generation of spontaneous polarization, especially in the ABX_3 perovskite configuration which has lower energy necessary for forming dipoles whose domain walls has low energies compared to other complex crystal systems as shown in the distorted cells of **Figure 11(b)**. The different kinds of crystal systems exert different types of lattice distortion as well as strain hence, the domain walls and width are influenced differently depending on the crystal system. Transitioning from one crystal symmetry to another also impacts the order parameter and the domain wall motion. Anisotropy is one important property found in ferroelectrics, which are dependent on the material as well as the crystal systems, certain

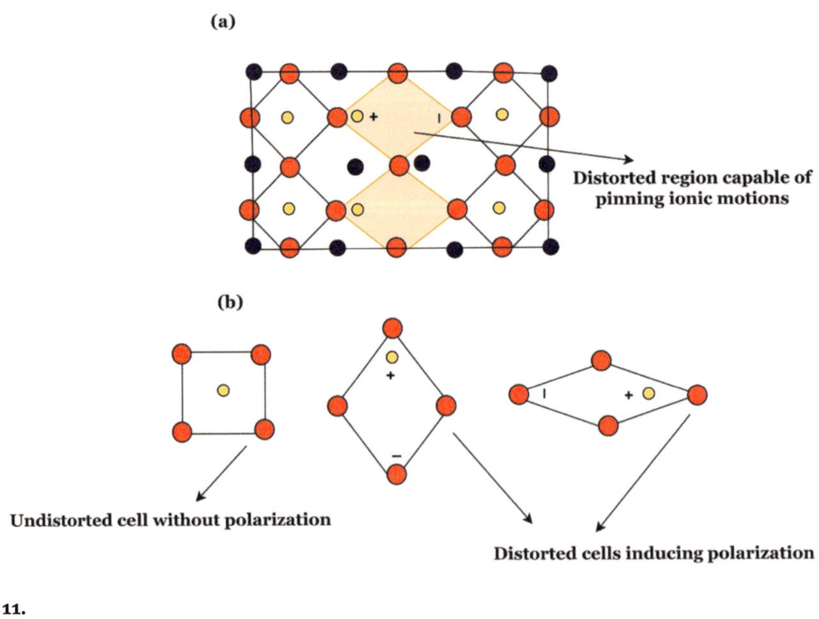

Figure 11.
Effects of lattices distortions as well as crystallography on domain wall.

crystallographic directions tend to have higher order parameters compared to adjacent directions [15]. This is possible due to the nature of domain wall motion, which differs ultimately depending on the direction of the applied field [39].

6.5 Grain and grain boundary

Generally, the physical properties, especially of the order parameters, are much more enhanced in the single crystalline phase than in the polycrystalline phase where there are so many micro grains. These grains have grain boundaries that are energetically high and therefore serve as a pinning center for the order parameters by preventing ionic motions, as shown in **Figure 12**. The shape, size, topology of the grain, and roughness of the grain boundaries are also parameters that greatly influence the domain wall motion, and these have been selected as variables for the engineering design of the domain wall configuration and mobility. The size and orientation of grains in a polycrystalline ferroelectric material can dictate the overall domain structure and domain wall behavior. Smaller grains may lead to more frequent encounters between domain walls and grain boundaries, affecting the material's macroscopic ferroelectric properties. Understanding the effects of grains and grain boundaries on ferroelectric domain walls is crucial for optimizing ferroelectric materials for various applications, including memory devices, sensors, actuators, and capacitors. Strategies to control grain size, orientation, and boundary properties can help tailor the performance of ferroelectric materials to specific technological requirements.

6.6 Domain interaction

Domain interaction is common in most ferroelectric materials and has a unique way of influencing domain mobility. The ferroelectric switching that contributes to the hysteresis loop can be greatly affected by domain interaction due to the impact of the interactions on the domain wall motion, which serves as an extrinsic contributor to the polarization switch. This interaction can be head-to-head, tail-to-tail, or

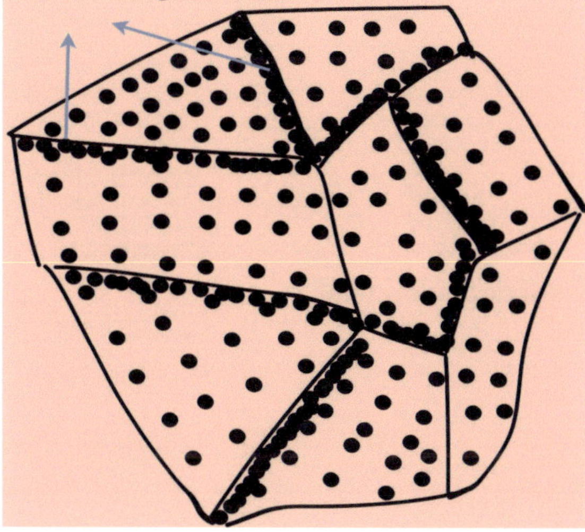

Figure 12.
Effects of grain boundaries on pinning down ionic motions.

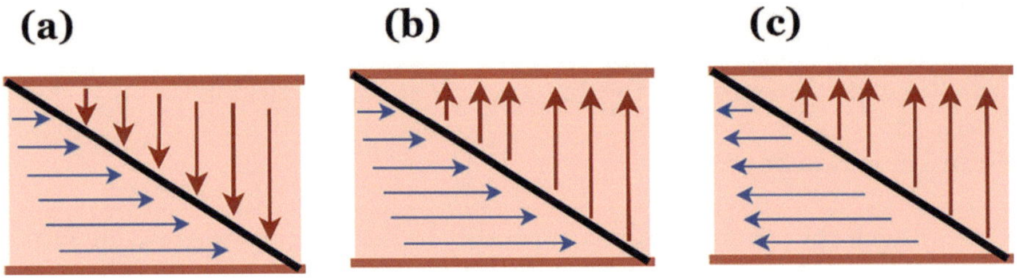

Figure 13.
Different configuration of domain interactions. (a) Head-Head, (b) Head-Tail, (c) Tail-Tail.

head-to-tail, as shown in **Figure 13**, and thus greatly influences the configuration of the domain wall in times of its topology, and even properties like conductivity. When the energy related to domain interactions is very low, then the associated energy of the domain wall can be minimal, allowing the domain wall to move to a lower applied field. However, when the interaction between domains is very strong, this can hamper domain wall motion and result in the domain wall's pinning. The strain and stress field related to domain wall interactions can also impact the width of the domain wall, which can either be narrowed or widened.

6.7 Dislocations

In ferroelectric materials, dislocation science is a common science. As is generally known, dislocations are line defects, either screw or edge (**Figure 14**), that frequently cause considerable amounts of strain fields to remain in a material, which can lead to localized lattice distortions. These tiny 2D flaws affect the domain's width, structure, and even motion when they contact with the walls. Improved order parameters is nearly impossible to achieve since greater energy dislocations can act as pinning centers for domain wall motion. The electrical and stress field close to the domain wall can also be changed by dislocations, which could change the field's characteristics and affect the polarization switch's functionality. Because

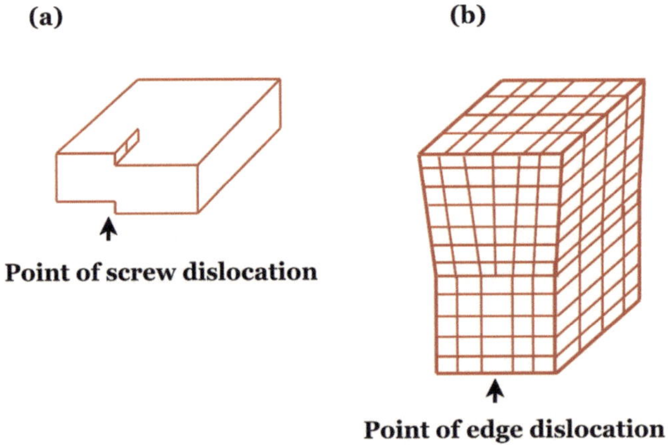

Figure 14.
Screw and edge dislocation. (a) Screw and (b) edge dislocation.

DOI: http://dx.doi.org/10.5772/intechopen.1006892

of their mechanical impacts, strain fields, and defect interactions, dislocations in ferroelectric materials are important factors that affect the behavior and characteristics of ferroelectric domain walls.

7. Domain and domain wall engineering in ferroelectrics

The major significance of domain and domain wall motion in ferroelectric materials is the contribution to piezoelectric and dielectric coefficients. This is true because these coefficients depend on the nature of the domain and its configuration. Several factors must be manipulated to engineer the domain and its wall to enhance the physical properties suitable for applications. Regarding the domain, the orientation of the polarization vectors and the sizes of the domain are often the manipulating parameters. Domain wall engineering involves the manipulation of adjustable parameters such as domain wall mobility and kinetics, topological structures, and conduction within the vicinity of the wall. Various techniques have been studied to manipulate these adjustable parameters to obtain enhancement in physical properties. The kind of engineering technique differs depending on the material: ferroelectric ceramic (polycrystal), single crystal, or thin film for which varying processing techniques are predominant depending on purpose and mode of application. Some of these domain and domain wall engineering techniques are discussed based on ceramics, single crystals, thin films, polymers, and composites.

7.1 Domain engineering

A good number of parameters have been considered valid for the engineering of the ferroelectric domain. These include the reorientation of the domain by poling which is mostly significant for piezoelectric studies. Poling involves the application of a very high electric field on the ferroelectric material which re-aligns all the polarization vectors of the adjacent domain to the direction of the applied electric field, to obtain a monodomain condition. The second technique for domain engineering is domain size reduction mostly induced by grain size reduction. Reduction of grain size can be achieved by manipulating the processing temperature. The grain size reduction strategy greatly reduces the long-range polarization order and thus increases the density of 90° domain walls.

7.2 Domain wall engineering

The engineering of the domain wall is dependent on the manipulation of parameters that greatly influence the domain wall's topological structure, conduction attributes, mobility, and kinetics. Some of the techniques such as grain size reduction and orientation strategies already discussed for the domain, also influence the domain wall. For instance, the density of the domain wall increases when the grain size reduces, and this greatly influences the topology and geometry, leading to a conduction in the wall. In addition, when the density of 90° walls increases, mobility and kinetics of the wall motion are facilitated thus contributing extrinsically to the electrical susceptibility [1].

7.3 Domain and domain wall engineering: ceramics

7.3.1 Mesoscopic chemical inhomogeneity

Chemical inhomogeneity in ceramic material arises when one or two constituent elements or an impurity dopant are distributed unevenly over the bulk ceramic, creating a concentration gradient in the bulk material. This inhomogeneity often occurs when highly volatile constituent elements or impurity dopants segregate when subjected to prolonged calcination and sintering time at very high temperatures and pressures. By adequately controlling the calcination temperature, it is possible to regulate the chemical inhomogeneity [1]. Chemical inhomogeneity is a major constraint influencing the topographical structure size, shape, and general configuration of the domain and domain wall in ferroelectric ceramics. This parameter can be adjusted using a suitable calcination and sintering temperature, pressure, and time. Although, it has been confirmed that prolonged annealing does not improve the chemical homogeneity [40], precursor technique, which involves the regulation of the processing parameters, helps in the facilitation of the mesoscopic chemical inhomogeneity. Domain and domain wall engineering by mesoscopic chemical inhomogeneity often results to the following:

i. Breakdown of the long-range domain-continuity into fine domain structures increasing domain wall density within a grain (formation of nano-size dense domain)

ii. Formation of nano polar inhomogeneous regions that exhibit non-linear piezoelectric response.

7.3.2 Grain-size effect

The average grain size of a polycrystalline ceramic shows a significant influence on the average density of the domain and domain wall. This is true because the grain boundary serves as a region of discontinuity of the domain, hence all domains within the grain are trapped within it. In terms of the bulk ceramic, the bulk average size of a ferroelectric ceramic must be >10 nm [41] to retain the ferroelectric properties. Generally, it has been reported that bulk size reduction leads to a decrease in the ferroelectric order of the ceramic material and thus disappears at <10 nm. This observation is based on bulk material. Contrary to the bulk size, the reduction in grain size increases the density of the 90° domain wall, which contributes significantly to the piezoelectric and dielectric response of the ferroelectric material. The grain size can be controlled during sintering by temperature manipulations.

7.4 Domain and domain wall engineering: single crystals

7.4.1 Chemical modification by doping

In most cases, ferroelectric single-crystal precursors are synthesized within the morphotropic phase boundary (MPB), where the coexistence of phases predominantly fosters ferroelectric-ferroelectric phase transition when the single crystal is subjected to external stimuli like temperature, electric field, pressure, or stress, etc. If we recall, domain and domain wall formation is a product of phase transition; hence, precursors synthesized within the MPB offer an opportunity for grown ferroelectric single crystals to be manipulated by the application of an external field to engender phase transition and, thus, manipulate the domain configuration through same. Compositional modification can also be achieved through acceptor and donor doping, which is significant in engineering the domain and domain wall. When a donor dopant is introduced into a B-site of a ferroelectric ABO_3 perovskite, say $PbTiO_3$, for instance, a donor dopant like Nb^{5+} influences the formation of Pb^{2+} vacancy at the "A" position. This vacancy tends to compensate for the dopant charges but becomes difficult to move on the application of an external field, and thus requires a large amount of energy to move the Pb^{2+} vacancy for instance from position (i) to position (ii) in **Figure 15(a)**. In this case, no complex defects are formed; hence, pinning of the domain wall is significantly impossible.

On the other hand, when an acceptor dopant like Fe^{3+} is introduced into the B-position, it engenders O^{2-} vacancies, and since the O^{2-} vacancies are located at the octahedral position, it has easy access to movement around this position; hence, less amount of energy is required to move the O^{2-} vacancy from position (i) to position (ii) as shown in **Figure 15(b)**. The problem with the O^{2-} vacancy is that it generates complex dipole defects that are often aligned in directions that pin the domain wall motion under applied fields until a particular threshold, then depinning sets in, and the complex defects dipoles will align in the same direction as the applied field, contributing to the order parameter. These defects engender by doping engineering results in the formation of hard and soft ferroelectrics in ceramic or single crystals whose domain wall is either clamped or unclamped for the donor and acceptor dopant, respectively. In undoped composition, the domain in $PbTiO_3$ behaves as though it is of a hard ferroelectric because the Pb^{2+} vacancies are compensated by O^{2-} vacancies, and therefore require large energy to shift the B-cation from one potential well to another in order to induce the hysteresis.

One effective method for controlling the characteristics of ferroelectric materials, such as domain wall behavior, is chemical doping. Comprehending the distinct impacts of dopants on domain walls is imperative for the enhancement of ferroelectric

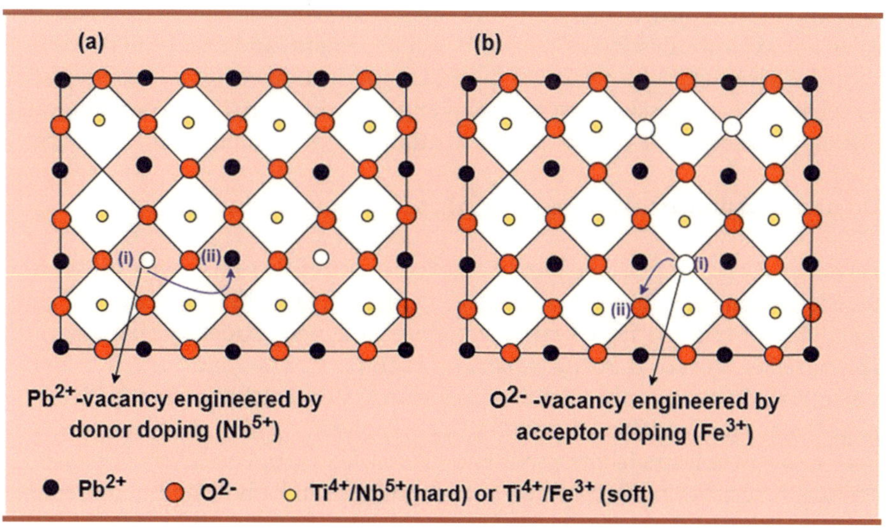

Figure 15.
Effect of the donor (a) and acceptor (b) dopant on domain and domain wall motion ($PbTiO_3$ system).

materials intended for a multitude of uses, encompassing memory devices, sensors, and actuators. The performance of ferroelectric materials can be precisely tuned by engineers to satisfy certain technical needs by controlling the kind, concentration, and distribution of dopants.

7.4.2 Poling

In ferroelectric single crystals, reorientation of the domain and the motion of the domain wall can be engineered by poling. The application of a high AC electric field with varying temperatures can be used to achieve poling in single crystals offering varying domain sizes. In other words, it is possible to engineer the domain and domain wall by poling to obtain fine domain sizes and thus enhance the piezoelectric properties. This technique has been used to enhance the piezoelectric properties of $BaTiO_3$ [42]. Poling contributes to the piezoelectric coefficients through the formation of 90° walls. Must ferroelectric materials exhibit anisotropic behavior having their physical properties different depending on the crystallographic directions. In this sense, when a single crystal is poled in a specific crystallographic direction, the piezoelectric and dielectric coefficients change depending on the direction of the applied field. For instance, in PZN-PT, the direction of spontaneous polarization is in the [111] directions, but poling the single crystal through [001], a giant piezoelectric and dielectric coefficient is observed. As it is with a single crystal, poling can also be achieved in ceramics where in both cases an alternating current is applied on the material plated with platinum or gold electrodes on either side which minimizes the effect of stray field and allows the domain to be poled by the applied field. Poling in ferroelectric can be achieved by applying an electrical field, while in ferromagnetic can be achieved by applying a magnetic field.

7.5 Domain and domain wall engineering: thin film

In thin films, domain and domain wall engineering can be achieved most importantly by the nature of the substrate material used. This substrate can engender phase transitions (for instance, $BiFeO_3$ thin film can transit from tetragonal to rhombohedral when an epitaxial strain substrate is used), enhancement, and change in the direction of the order parameter like polarization and magnetization vectors. The application of either tensile or compressive stress can also engineer the domain and domain wall in thin films. Others include poling, chemical modification, temperature, light, etc.

7.6 Domain and domain wall engineering: polymers and composites

Ferroelectric polymers, copolymers, and composites also exist with well-defined hysteresis showing enhanced piezoelectric, dielectric, and electromechanical coefficients. The existence of ferroelastic, ferromagnetic, and ferroelastic behaviors can also be traced in most of these materials. An example of a polymer with well-defined ferroelectric behavior is polyvinylidene fluoride (PVDF). Aside from the PVDF homopolymer, the Co-polymer, polyureas, and polyamide [43], have been proven to show significant ferroelectric properties. Ferroelectric polymers are not pure crystalline materials but a mixture of both crystalline and amorphous nanoregions, which in addition to changes to crystal symmetry during phase transition, also show morphological changes in the crystal-amorphous nanoregions that greatly influence their ferroelectric properties [44, 45]. At various suitable compositions, copolymers of PVDF with trifluoro ethylene have shown tendencies of temperature-dependent ferroelectric-paraelectric phase transition with clear Tc [44] which makes them not only suitable piezoelectric materials but also potential pyroelectric materials with high-temperature stability [46], and thus paving the way for multifunctionality.

In this ferroelectric polymer material, the domain and domain wall can be engineered by the previously mentioned techniques, including compositional modification and poling. Engineering the ratio of crystal-amorphous nanoregions within the ferroelectric polymer can also significantly influence the ferroelastic and ferroelectric domains. Introducing inorganic ferromagnetic oxides can also help induce ferromagnetism and thus engineer the nature and configuration of the domain wall. Li et al. [45] have reported the possibility of surface engineering through the embedment of nano-ferroelectric inorganic oxides into the matrix of these polymers to further enhance their ferroelectric by inducing other ferroelectric properties and thus fostering multiferroicity in ferroelectric polymers. Ferroelectric composites are also being studied recently and how domain and domain wall engineering can help enhance the physical properties. All the mentioned techniques can be used to engineer domain and domain wall motion in ferroelectric composites.

8. Important applications of domain wall motions

As we shall see in this section, domain wall motion and the dynamics of the domain itself have been explored for a lot of applications in memory storage, actuators, transducers, sensors, and capacitors.

8.1 Hysteresis in ferroelectricity: storage memory and capacitors

The idea behind the improvement in memory capacity stems from the idea that domain wall motion contributes significantly to ferroelectric hysteresis. In computer technology, information in electric signals is stored as "0" or "1". In the ferroelectric memory device, the polarization vector can be poled from one direction, left, into the other, right. The direction left can be tagged "0" and the direction right can be tagged "1" or vise vasa, such that every electrical signal coded with information that comes in the forms of "0" or "1" can pole the polarization vectors at different domains allowing for the storage of such information coded with the electrical signal (**Figure 16**). Domain wall motion helps a lot in enhancing the capacity of storage memory devices, and therefore, controlling the motion by pining the domain wall can help archive the specified behavior for storage memory. Domain wall pinning can be a geometrical or non-geometrical approach that can be used to control the stored information.

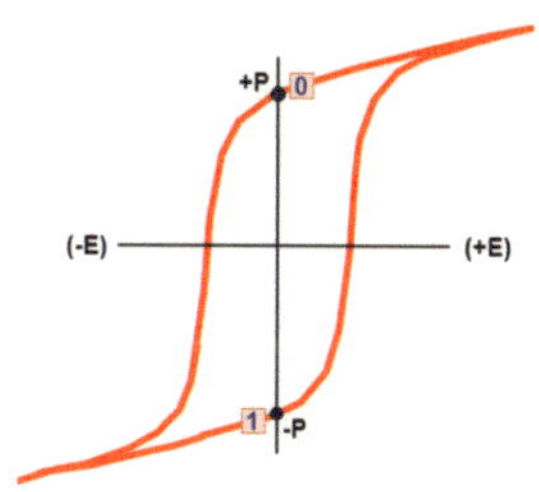

Figure 16.
Coding of '0' and '1' in ferroic material.

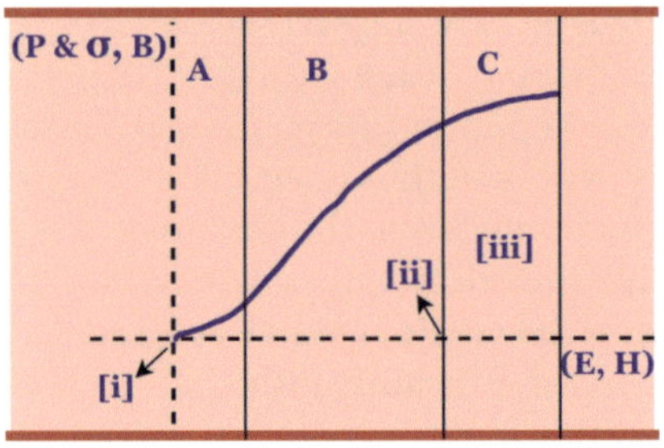

Figure 17.
Non-linear response of 90° domain wall motion.

In domain wall memory, information can be stored first by writing (Nucleation of the domain), addressing the domain (moving the domain wall), and finally reading the domain wall [5]. All these processes follow the hysteresis loop. Both 180° and 90° domain walls contribute majorly to ferroelectric hysteresis and, thus, the memory storage properties. Again, the 90° domain wall contributes significantly to the dielectric constant, which is a parameter that is significant to the capacitance of a capacitor and therefore enhances the number of charges that can be stored on a capacitor. This is very significant in modern-day technology as almost all devices have in-build capacitors.

8.2 Piezoelectricity: actuators, transducers, and sensors

It has been established that the 90° wall motion contributes to piezoelectric coefficients, whereas the 180° wall contributes to the higher order electrostrictive coefficients. The contribution of 90° domain wall motion to piezoelectric coefficients is crucial for actuators and transducer applications. The region labeled A in **Figure 17** represents the non-linear region that responds significantly to 90° domain wall motion via enhanced vibrational motion and electric field-induced strain, which are thus harvested for actuating and transducing functionality. The working principles of transducers depend on the quality of the electromechanical coupling factors, piezoelectric coefficients, and dielectric coefficients. Transducers can be used as biosensors, thickness detectors, nano balances, and medical imaging, which employ the functionality of a piezoelectric element. Piezo sensors have also been developed, which depend significantly on piezoelectric coefficients, another coefficient dependent on 90° domain wall motion. Regions B and C are of the coercive and saturation field effects and mostly break the piezoelectric response by introducing ferroic hysteresis.

9. Summary

Domain and domain walls have been considered twin defects that arise in a ferroic material during phase transition from high symmetric phases to a lower one or between two lower symmetric phases. As we have seen in this chapter, domain and domain wall motion play a crucial role in the major advances in electronic device fabrication, and therefore, their formation and engineering are highly significant for industrial applications. The first section of this chapter introduces domain and domain wall, followed by the techniques and reason for their formation and how they can be visualized. Geometrically, we understood that domain walls can form 180 and non-180-degree domain walls, whereas in some cases, the domain walls are either straight lines or curves with the ability to conduct electrical charges. The Ising, Bloch, and Neel domains classified based on the nature of rotation of the order parameter about the planes of the domain wall have also been covered in this chapter. A large section of the chapter is dedicated to domain wall motion and its contribution to the individual ferroic classes. This is very significant as it leads to the various engineering techniques that depend on the motion of the domain wall to engender property change by changing the configuration of the domain and the domain wall. The engineering techniques to be deployed for archiving this purpose include poling, grain-size effect, compositional modification, and so on, and they are material dependent. Lastly, the chapter provides a vivid communication on the various applications of domain walls, especially 90° domain walls, which contribute mostly to the piezoelectric coefficient and dielectric constant, making it accessible engineering properties for fabrication actuating, transducing, sensing devices as well as capacitors with very high capacitance.

Acknowledgements

The authors thank FAPESP (Proc. 2017/13769-1 and 2022/05578-0), CNPq (Proc – 307783/2020-0), and FINEP for the support. The authors also thank the Federal University of Sao Paulo for providing the core research facilities—NAPCEM.

Author details

Shedrach Yakubu
Group of Multifunctional Materials and Sensors—GMMS, Department of Science and Technology, Federal University of São Paulo, Brazil

*Address all correspondence to: shedrachyakubu1@gmail.com

References

[1] Thong HC, Li Z, Lu J-T, Li C-B-W, Liu Y-X, Sun Q, et al. Domain engineering in bulk ferroelectric ceramics via mesoscopic chemical inhomogeneity. Advanced Science. 2022;**9**(17):1-9. DOI: 10.1002/advs.202200998

[2] Schönecker AJ, Nicolai M, Schlenkrich F, Henschke T, Uhlig S. Investigation of domain switching of PZT and KNN—Based ceramics. In: Proceedings of the International Symposium on Applications of Ferroelectrics and International Symposium on Piezoresponse Force Microscopy and Nanoscale Phenomena in Polar Materials, ISAF/PFM 2011. Canada: IEEE; 2011. pp. 34-37. DOI: 10.1109/ISAF.2011.6014107

[3] Liu Y, Kelley KP, Funakubo H, Kalinin SV, Ziatdinov M. Exploring physics of ferroelectric domain walls in real time: Deep learning enabled scanning probe microscopy. Advanced Science. 2022;**9**(31):1-10. DOI: 10.1002/advs.202203957

[4] Venkat G, Allwood DA, Hayward TJ. Magnetic domain walls: Types, processes and applications. Institute of Physics. 2024;**57**(6):1-32. DOI: 10.1088/1361-6463/ad0568

[5] Kumar D, Jin T, Al Risi S, Sbiaa R, Lew WS, Piramanayagam SN. Domain Wall motion control for racetrack memory applications. IEEE Transactions on Magnetics. 2019;**55**(3):1-8. DOI: 10.1109/TMAG.2018.2876622

[6] Tagantsev AK, Cross LE, Fousek J. Domains in Ferroic Crystals and Thin Films. 1st ed. New York, Dordrecht, Heidelberg, London: Springer; 2010. DOI: 10.1007/978-1-4419-1417-0

[7] Grünebohm A, Marathe M, Khachaturyan R, Schiedung R, Lupascu DC, Shvartsman VV. Interplay of domain structure and phase transitions: Theory, experiment and functionality. Journal of Physics: Condensed Matter. 2022;**34**:1-38. DOI: 10.1088/1361-648X/ac3607

[8] Evans DM, Garcia V, Meier D, Bibes M. Domains and domain walls in multiferroics. Physical Sciences Reviews. 2020;**5**(9):1-23. DOI: 10.1515/psr-2019-0067

[9] Shilo D, Ravichandran G, Bhattacharya K. Investigation of twin-wall structure at the nanometre scale using atomic force microscopy. Nature Materials. 2004;**3**(7):453-457. DOI: 10.1038/nmat1151

[10] Kish LL, Thaler A, Lee M, Zakrzewski AV, Reig-i-Plessis D, Wolin BA, et al. Domain wall patterning and giant response functions in ferrimagnetic spinels. Advanced Science. 2021;**8**(23):1-13. DOI: 10.1002/advs.202101402

[11] Mukherjee D. Growth and Characterization of Epitaxial Thin Films and Multiferroic Heterostructures of Ferromagnetic and Ferroelectric Materials. USA: College of Arts and Sciences University of South Florida; 2010

[12] Schultheiß J, Picht G, Wang J, Genenko YA, Chen LQ, Daniels JE, et al. Ferroelectric Polycrystals: Structural and Microstructural Levers for Property-Engineering Via Domain-Wall Dynamics. USA: Elsevier Ltd; 2023. DOI: 10.1016/j.pmatsci.2023.101101

[13] Li CBW, Thong HC, Liu YX, Bi K, Fu Z, Wang K. Thermally induced

domain reconfiguration in ferroelectric alkaline niobate. Advanced Functional Materials. 2022;**32**(38):1-9. DOI: 10.1002/adfm.202204421

[14] Uchino K. Ferroelectric Devices. 2nd ed. Boca Raton, London, New York: Taylor & Francis Group, LLC; 2010. pp. 1-349

[15] Schneider D, Jo W, Rödel J, Rytz D, Granzow T. Anisotropy of ferroelectric behavior of (1 - X)Bi1/2Na 1/2TiO3-xBaTiO3 single crystals across the morphotropic phase boundary. Journal of Applied Physics. 2014;**116**(4):1-9. DOI: 10.1063/1.4891529

[16] Wersing W, Lubitz K, Mohaupt J. Anisotropic piezoelectric effect in modified PbTi03 ceramics. In: IEEE Transactions on Ultrasonics, Ferroelectrics, and Frequency Control. Vol. 36. Canada: IEEE; 1989. pp. 424-433

[17] Kittel C. Introduction to Solid State Physics. 8th ed. United States of America: John Wiley & Sons, Inc; 2005

[18] Gersten JI, Smith FW. The Physics and Chemistry of Materials. United States of America: Wiley; 2001

[19] Cross LE. Antiferroelectric-ferroelectric switching in a sample 'Kittel' antiferroelectric. Journal of the Physical Society of Japan. 1967;**23**(1):77-82

[20] Kittel C. Physical theory of ferromagnetic domains. Reviews of Modern Physics. 1949;**21**(4):541-583

[21] Chen IW. Dielectric and ferroelectric ceramics: Interfaces. In: Encyclopedia of Materials: Science and Technology. Greenwich: Mannial Pub. Co.; 2001. pp. 2152-2157. DOI: 10.1016/b0-08-043152-6/00388-0

[22] Bhide VG, Bapat NJ. Interferometric study of domain structure in barium titanate. Physica. 1961;**27**(6):531-540. DOI: 10.1016/0031-8914(61)90070-2

[23] Vasudevan RK, Morozovska AN, Eliseev EA, Britson J, Yang J-C, Chu Y-H, et al. Domain wall geometry controls conduction in ferroelectrics. Nano Letters. 2012;**12**(11):5524-5531. DOI: 10.1021/nl302382k

[24] Bednyakov PS, Sturman BI, Sluka T, Tagantsev AK, Yudin PV. Physics and applications of charged domain walls. Nature Publishing Group. 2018;**65**:1-11. DOI: 10.1038/s41524-018-0121-8

[25] Ummelen F, Swagten H, Koopmans B. Racetrack memory based on in-plane-field controlled domain-wall pinning. Scientific Reports. 2017;**7**(1):1-8. DOI: 10.1038/s41598-017-00837-x

[26] Mbarki R, Borvayeh L. Investigation of the flexoelectric coupling effect on the 180° domain wall structure and interaction with defects. Journal of Material Science & Engineering. 2016;**5**(4):1-6. DOI: 10.4172/2169-0022.1000264

[27] Lee D, Behera RK, Pingping W, Haixuan X, Li YL, Sinnott SB, et al. Mixed Bloch-Néel-Ising character of 180°ferroelectric domain walls. Physical Review B: Condensed Matter and Materials Physics. 2009;**80**(6):1-4. DOI: 10.1103/PhysRevB.80.060102

[28] Padilla J, Zhong W, Vanderbilt D. First-principles investigation of 180° domain walls in BaTiO3. Physical Review B: Condensed Matter. 1996;**53**(10):5969-5973

[29] Liu Z, Wang H, Li M, Tao L, Paudel TR, Hongyang Y, et al. In-plane charged domain walls with memristive behaviour in a ferroelectric film. Nature. 2023;**613**(7945):656-661. DOI: 10.1038/s41586-022-05503-5

[30] Shur VY. Fast polarization reversal process: Evolution of ferroelectric domain structure in thin films. In: Paz de Araujo CA, Scott JF, Tailor GW, editors. Ferroelectricity and Related Phenomenon Series. Vol. 10. London: Gordon and Breach Science Publ; 1996. pp. 153-192. Available from: https://www.researchgate.net/publication/239556517

[31] Marincel DM, Zhang H, Jesse S, Belianinov A, Okatan MB, Kalinin SV, et al. Domain wall motion across various grain boundaries in ferroelectric thin films. Journal of the American Ceramic Society. 2015;**98**(6):1848-1857. DOI: 10.1111/jace.13535

[32] Damjanovic D. Ferroelectric, Dielectric and Piezoelectric Properties of Ferroelectric Thin Films and Ceramics. Bristol: IOP publishing Ltd.; 1998

[33] Herbiet R, Robels U, Dederichs H, Arlt G. Domain wall and volume contributions to material properties of PZT ceramics. Ferroelectrics. 1989;**98**(1):107-121. DOI: 10.1080/00150198908217575

[34] Lente MH, Eiras JA. Interrelationship between self-heating and ferroelectric properties in PZT ceramics during polarization reorientation. Journal of Physics: Condensed Matter. 2000;**12**:5939-5950

[35] Yuxiang H, Pan Y, Wang Z, Lin T, Gao Y, Luo B, et al. Lattice distortion induced internal electric field in TiO2 photoelectrode for efficient charge separation and transfer. Nature Communications. 2020;**11**(1):1-10. DOI: 10.1038/s41467-020-15993-4

[36] Dangić D, Murray ED, Fahy S, Savić I. Structural and thermal transport properties of ferroelectric domain walls in GeTe from first principles. Physical Review B. 2020;**101**(18):1-8. DOI: 10.1103/PhysRevB.101.184110

[37] Shi Q, Eric P, Xiaoxing C, Natalya F, Ren-Ci P, Abel F, et al. The role of lattice dynamics in ferroelectric switching. Nature Communications. 2022;**13**(1):1-10. DOI: 10.1038/s41467-022-28622-z

[38] Hayashida T, Uemura Y, Kimura K, Matsuoka S, Hagihala M, Hirose S, et al. Phase transition and domain formation in ferroaxial crystals. Physical Review Materials. 2021;**5**(12):1-17. DOI: 10.1103/PhysRevMaterials.5.124409

[39] Zhuo F, Zhou X, Gao S, Höfling M, Dietrich F, Groszewicz PB, et al. Anisotropic dislocation-domain wall interactions in ferroelectrics. Nature Communications. 2022;**13**(1):1-11. DOI: 10.1038/s41467-022-34304-7

[40] Wang Y, Damjanovic D, Klein N, Setter N. High-temperature instability of Li- and Ta-modified (K,Na)NbO3 piezoceramics. Journal of the American Ceramic Society. 2008;**91**(6):1962-1970. DOI: 10.1111/j.1551-2916.2008.02392.x

[41] Buscaglia V, Randall CA. Size and Scaling Effects in Barium Titanate. An Overview. Vol. 40, No. 11. Elsevier; 2020. pp. 3744-3758. DOI: 10.1016/j.jeurceramsoc.2020.01.021

[42] Wada S, Yako K, Yokoo K, Kakemoto H, Tsurumi T. Domain wall engineering in barium titanate single crystals for enhanced piezoelectric properties. Ferroelectrics. 2006;**334**(1):17-27. DOI: 10.1080/00150190600689647

[43] Li Q, Wang Q. Ferroelectric polymers and their energy-related applications. Macromolecular Chemistry and Physics. 2016;**217**(11):1228-1244. DOI: 10.1002/macp.201500503

[44] Hafner J, Benaglia S, Richheimer F, Teuschel M, Maier FJ, Werner A, et al. Multi-scale characterisation of a ferroelectric polymer reveals the emergence of a morphological phase transition driven by temperature. Nature Communications. 2021;**12**(1):1-9. DOI: 10.1038/s41467-020-20407-6

[45] Li J, Liu X, Gu J, Chen D, Feng Y, Zhang W, et al. Investigation on tunable ferroelectricity of polymer-based composites via changing the surface state of ferroelectric nanosheets. Materials Today Energy. Mar 2023;**32**:1-9. DOI: 10.1016/j.mtener.2022.101236

[46] Fang L, Tao Y, Wang C, Dai M, Huang G, Sun J, et al. Resveratrol-based fluorinated materials with high thermostability and good dielectric properties at high frequency. ACS Sustainable Chemistry & Engineering. 2020;**8**(45):16905-16911. DOI: 10.1021/acssuschemeng.0c06061

Chapter 3

Structure and Symmetry of Ferroelectric $BaTiO_3$-Based Ceramics

Ahmed Gadelmawla

Abstract

The physical property of any material is directly correlated to its crystal symmetry. Therefore, a better understanding of crystal structure of ferroelectrics leads to a better understanding of their properties and future application under external fields. In this chapter, an investigation of $BaTiO_3$-based (BCZT) crystal structure and the changes in the structure as a function of chemical composition, and temperature, provide a better understanding of the material's properties (e.g. spontaneous polarization). In addition, the effect of Ca and Zr substitution on BCZT local structure and phase boundaries are reported for both bulk and single crystal, using several structural characterization methods, e.g., XRD and XFH.

Keywords: BCZT, X-ray fluorescence holography, X-ray diffraction, single crystal, perovskites ferroelectric materials, crystallography

1. Introduction

Ferroelectric materials are widely utilized in various industries, including medical, aerospace, transportation, and energy sectors. They are crucial in everyday applications, particularly in energy generation via the piezoelectric behavior of ceramics. This feature is especially significant for energy harvesting systems that provide localized, wireless energy access for unattended wireless sensors. These sensors are used for monitoring pipe or bridge corrosion, detecting temperature, humidity, dust, smog, and gases (like ozone), medical sensor arrays (e.g., on shoes for diagnosing Parkinson's disease), and remote telecommunications [1–7]. To examine the properties of lead-free ferroelectric materials and their behavior under different conditions, it has been observed that the physical properties of these crystals are linked to their symmetry. Franz Neumann first articulated the fundamental principle of the structure-property relationship in 1885 [8], known as *Neumann's Principle*. It can be stated as follows:

> "*The symmetry elements of any physical property of a crystal must include the symmetry elements of the point group of the crystal.*"

This means that a physical property can exhibit higher symmetry than the crystal itself but cannot have lower symmetry. In essence, the point group of the property must be a super-group of the crystal's point group. The point group of a crystal identifies the symmetrical elements in its structure. Crystallography categorizes all crystal families into seven crystal systems and the 14 Bravais lattices (including the face centering): cubic, hexagonal, tetragonal, rhombohedral, orthorhombic, monoclinic, and triclinic (see **Figure 1**). Crystal symmetry is further classified into 14 Bravais lattices and 32 crystal classes, known as point groups. Incorporating centering and translation elements results in 230 possible space groups, which encompass all information about the structure's symmetry (**Table 1**) [9, 10].

Crystal physics focuses on the macroscopic properties of crystals in equilibrium, referred to as reversible properties. This means that when an external force field is removed, any field-induced changes in the crystal are completely reversed. In describing physical properties, tensors are used to represent the direction and magnitude of these reversible properties, linking an applied field to its effects on the crystal [10]. Typical tensor properties that will be covered in this chapter are dielectric properties (*i.e.*, ferroelectric and piezoelectricity).

Macroscopic properties such as volume expansion, dielectric properties, and mechanical response change under external fields like thermal, electrical, and mechanical stresses. These changes can be explained by investigating microscopic changes in the crystal structure. *In situ,* crystal structure studies under external fields are commonly used to correlate microscopic changes with macroscopic properties. Techniques such as piezoelectric force microscopy, *in situ* XRD, Raman spectroscopy,

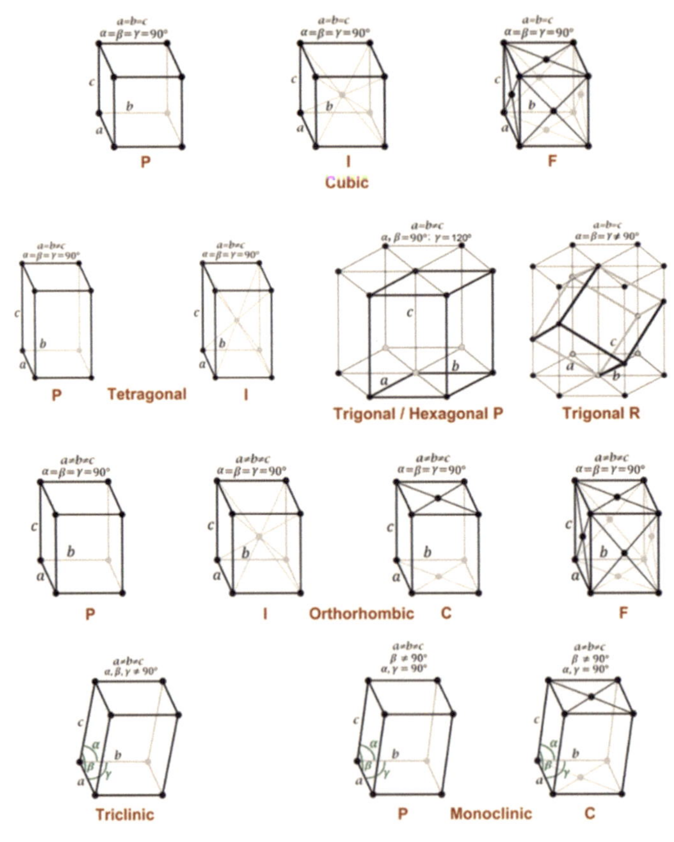

Figure 1.
The 14 Bravais lattices redrawn after [9].

Symbol	Symmetry element	Symmetry operation
P	Primitive unit cell	Basic unit cell with lattice coordinate at 0, 0, 0
A	A-face centered	Unit cell centered in the A-face with lattice coordinate at 0, 0, 0; 0, 0.5, 0.5
B	B-face centered	Unit cell centered in the B-face with lattice coordinate at 0, 0, 0; 0.5, 0, 0.5
C	C-face centered	Unit cell centered in the C-face with lattice coordinate at 0, 0, 0; 0.5, 0.5, 0
F	All faces centered	Unit cell with all faces centered, A, B, C
I	Body centered	Unit cell where the same atom located at the corners and the center, with lattice coordinates 0, 0, 0; 0.5, 0.5, 0.5
R	Rhombohedral centered	Primitive unit cell (described with rhombohedral axes)
2, 3, 4, 6	Number of fold rotation axis	Counter-clockwise rotation of 360/n degrees around the axis
$\bar{1}, \bar{2}, \bar{3}, \bar{4}, \bar{6}$,	Rotoinversion axis and inversion center	Counter-clockwise rotation of 360/n degrees around the axis, followed by inversion through the point on the axis
2_p,3_p,4_p,6_p	Number of fold screw axis	Right-handed screw rotation of 360/n degrees around the axis, with screw vector (pitch) (p/n)
m	Reflection plane, mirror plane	Reflection through the plane
a, b, c	Axial glide plane	Glide reflection through the plane, with glide vector 0.5**a**; 0.5**b**; 0.5**c**
d	Diamond glide plane	Glide reflection through the plane, with glide vector 0.25(**a** ± **b**); 0.25(**a** ± **c**); 0.25(**b** ± **c**); etc.
n	Diagonal glide plane	Glide reflection through the plane, with glide vector 0.5(**a** ± **b**); 0.5 (**a** ± **c**); 0.5(**b** ± **c**); etc.

In crystallography, the bold a, b, c means the direction of transition.

Table 1.
Space group symbols definition for symmetry elements and operations [9].

and TEM illustrate crystal structure changes under different fields, including temperature, stress, and direct current.

Among the 32 crystal classes, 21 are non-centrosymmetric, meaning they lack rotoinversion symmetry operations (i.e., translating from x, y, z to $-x, -y, -z$ makes no difference). The remaining 11 centrosymmetric classes are nonpolar and exhibit no ferroelectric effect. Although non-centrosymmetric classes lack roto inversion, only one class is nonpolar due to the potential inclusion of multiple rotation axes or a mirror plane perpendicular to a rotation axis, making all cubic classes nonpolar. Under mechanical stress, 10 non-centrosymmetric classes can be polarized, while the other 10 (i.e., $1, 2, m, mm2, 3, 3m, 4, 4mm, 6, 6mm$) are spontaneously polarized, encompassing piezoelectric, pyroelectric, and ferroelectric materials with reversible polarization. Ferroelectric materials exhibit reversible spontaneous electric polarization and strain. Of the 230 space groups, only 68 belong to the ten polar point groups that are classified as ferroelectrics (see **Table 2**).

One of the most famous ferroelectrics is perovskite. Perovskite $CaTiO_3$, named after the Russian mineralogist Lev Perovski, is a class of materials with a distinctive crystal structure first discovered in the Ural Mountains of Russia in 1839 by Gustav

Point group	Symmetry classes	Space group
C1	1	$P1$
C2	2	$P2, P2_1, C2$
C1*h*	m	Pm, Pc, Cm, Cc
C2*v*	$mm2$	$Pmm2, Pmc2_1, Pcc2, Pma2, Pca2_1, Pnc2, Pmn2_1, Pba2, Pna2_1, Pnn2, Cmm2, Cmc2_1, Ccc2, Amm2, Abm2, Ama2, Aba2, Fmm2, Fdd2, Imm2, Iba2, Ima2$
C4	4	$P4, P4_1, P42, P4_3, I4, I4_1$
C4*v*	$4mm$	$P4mm, P4bm, P4_2cm, P4_2nm, P4cc, P4nc, P4_2mc, P4_2bc, I4mm, I4cm, I4_1md, I4_1cd$
C3	3	$P3, P31, P32, R3$
C3*v*	$3m$	$P3m1, P31m, P3c1, P31c, R3m, R3c$
C6	6	$P6, P6_1, P6_5, P6_2, P6_4, P6_3$
C6*v*	$6mm$	$P6mm, P6cc, P6_3cm, P6_3mc$

Table 2.
Ferroelectric space groups belong to the 10 polar point groups.

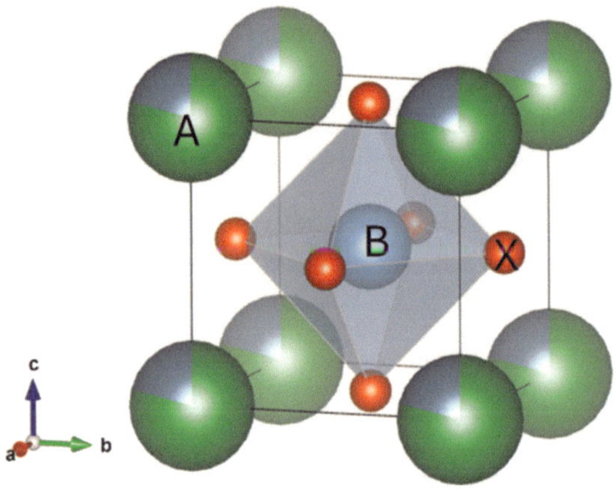

Figure 2.
The perovskite ABX_3 structure ($Pm\bar{3}m$).

Rose [11]. The perovskite structure is characterized by the general formula ABX_3, where 'A' and 'B' are cations of different sizes, and 'X' is an anion, typically oxygen (**Figure 2**). In this structure, the 'A' cation is larger and occupies the corners of a cubic unit cell, while the smaller 'B' cation sits at the center, surrounded by an octahedron of 'X' anions. This arrangement forms a highly symmetrical and versatile framework that can accommodate a wide variety of elements, resulting in diverse chemical compositions and physical properties. In its ideal cubic form, the perovskite structure exhibits a high degree of symmetry, classified under the space group $Pm\bar{3}m$.

Goldschmidt made significant contributions to determine the perovskite structures through his formulation of the Goldschmidt tolerance factor in the 1920s. This factor helps predict the stability of the perovskite structure based on the ionic radii of the constituent elements. The tolerance factor (t) is given by the formula:

$$t = \frac{r_A + r_X}{\sqrt{2}(r_B + r_X)}$$

Where r_A, r_B and r_X are the ionic radii of the A cation, B cation, and X anion, respectively. The value of the t can predict the crystal symmetry of the perovskite material. The perovskite structure typically forms a cubic phase when $t = 1$. Whereas in case of $t \neq 1$, The structure tends to distort to lower symmetry phases such as tetragonal, orthorhombic, or rhombohedral [12].

This versatility has made perovskites crucial in the development of ferroelectric materials. The discovery of ferroelectricity in barium titanate ($BaTiO_3$) during the 1940s marked a significant milestone. Barium titanate, with its perovskite structure, exhibited spontaneous electric polarization that could be reversed by the application of an external electric field, a property that defines ferroelectric materials [1–5]. Among the possible barium titanate-based polycrystalline $Ba_{1-x}Ca_xTi_{1-y}Zr_yO_3$ (BCZT) has been found to display an exceptional piezoelectric coefficient d_{33} ~620 pC/N, whereas the predicted single crystal may reach 2000 pC/N [13], making it particularly attractive for both actuation and sensing applications. In this chapter, *in situ* temperature dependence XRD was used to investigate the changes in symmetry of both BCZT single crystal and bulk as a function of temperature, and the effect of Ca and Zr on the local structure was reported [14–17]. These data are crucial for understanding the ferroelectric behavior of lead-free perovskites and their properties. This work provides significant insights into the operational conditions of lead-free ferroelectrics, aiding in their application for future technologies.

2. The crystal structure of $BaTiO_3$-based material

Barium titanate-based materials typically undergo three-phase transitions during heating from low temperatures, moving through the following phases: rhombohedral (R), orthorhombic (O), tetragonal (T), and cubic (C) [18–21]. These phase transitions can be altered both chemically through aliovalent and isovalent substitutions, as well as microstructurally by changing the grain size. Additionally, externally applied stress can also shift these transitions [22–25]. Depending on the applied temperatures, eight space groups of crystal structures have been reported for $BaTiO_3$ (BTO), *i.e.*, $Pm\bar{3}m$ (a primitive cubic unit cell with two mirrors and 3-fold rotoinversion axis), *P4mm*, $P4/mmm$ (two primitive tetragonal unit cells with 4-fold axis and two mirrors, another possibility with a perpendicular mirror on the 4-fold axis), *Pmm2*, *Amm2*, $C222_1$ (three orthorhombic unit cell with primitive, *A*-face, and *C*-face centered, combined with two mirrors and 2-fold axis), *R3m* (a primitive trigonal unit cell with rhombohedral 3-fold axis and a mirror) and even the hexagonal polymorph $P6_3/mmc$ (a primitive hexagonal unit cell with 6_3-fold screw axis, perpendicular on a mirror with glid plane in the **c** direction) [26, 27]. Importantly, by adjusting the A- and B-site atomic radii and the oxidation state of the substituting element, the phase existence and stability at ambient conditions can be significantly tailored [28, 29]. For example, doping calcium (Ca) on the A-site of barium titanate (BTO) reduces the dielectric constant and spontaneous polarization, as calcium titanate ($CaTiO_3$) is not ferroelectric and transitions from tetragonal to cubic at a very high Curie temperature (T_C) [30, 31]. This doping increases the T_C with an increasing concentration of Ca up to 20 mol% [32]. In contrast, adding zirconium (Zr) on the B-site of BTO suppresses the TC, leading to an increase in the maximum relative permittivity and a pronounced diffuse phase transition behavior for Zr contents greater than approximately 20 mol%, consistent with a transition from the ferroelectric to the relaxor state [33, 34]. Therefore, co-substituting Ca and Zr has been observed to result in a relatively higher T_C than BZT and a higher dielectric constant than BCT simultaneously [35]. The influence of Ca and Zr substitutions on the A- and B-sites on the crystal structure and phase transitions was previously reported [18, 21, 36] and can be rationalized by considering the Shannon radii: Ca (1.34 Å), Ba (1.61 Å), Ti (0.61 Å) and Zr (0.72 Å).

2.1 *In situ* temperature-dependent X-ray diffraction study of ferroelectric single crystal BCZT

The average structure of BCZT (barium calcium zirconium titanate) solid solution has been extensively studied and reported in previous research for both powder and bulk samples [18, 21, 37, 38]. For example, Haugen et al. observed three phases (rhombohedral $R3m$, tetragonal $P4mm$, and cubic $Pm\bar{3}m$) within a temperature range of −100–152°C using Rietveld refinement for BCZT50, identifying three-phase transition steps at −96°C, 34°C, and 90°C between R, R + T, T, and C phases, respectively [37]. Later, Keeble et al. proposed the existence of an interleaving orthorhombic phase ($Amm2$) between the rhombohedral and tetragonal phases after investigating the full range of BCZT0–100 from 80 to 450 K [36]. However, these studies have not mentioned the crystal structure and phase transitions of BCZT single crystals. Single crystals are essential for a more precise analysis of the symmetry, phase transitions, static configurational displacement, and thermal vibrations of each ion. Parameters that are not easily obtained from powder diffraction data, especially the anisotropic displacement model of the ions [14].

Several earlier studies have successfully grown BCZT single crystals [39–43]. However, only Liu et al. have reported a high piezoelectric coefficient d_{33} of 232 pC/N at room temperature for BCZT single crystal ($Ba_{0.798}Ca_{0.202}Zr_{0.006}Ti_{0.994}O_3$) [44]. The relatively high d_{33} makes Liu's crystal the best candidate for structural investigation. This single crystal was grown using the spontaneous nucleation method; additional details about the synthesis, crystal growth, and physical properties are available elsewhere [44]. Single crystal X-ray diffraction data were collected at −103°C, −53°C, 25°C, 47°C, and 107°C to investigate all three ferroelectric phases below the Curie temperature; additional details about the instrument can be found elsewhere [14].

A portion of the single crystal was selected from the mother sample for X-ray crystal structure determinations (see **Figure 3**). According to the dielectric data reported by Liu et al., phase transitions occurred around −75°C, 6°C, and 127°C for R-O, O-T, and T-C transitions, respectively [14]. X-ray diffraction data were collected at −103°C, −53°C, 25°C, 47°C, and 107°C to investigate all three ferroelectric phases below the Curie temperature. Temperature control was achieved using an Oxford Cryosystems Series 700 cryostream cooler, maintaining a gas stability of 0.1 K from −193–127°C. Intensity data were collected using MoKα radiation (λ = 0.71073 Å) on a Bruker Kappa PHOTON 2 IμS Duo diffractometer with QUAZAR focusing Montel optics. Data were corrected for Lorentz and polarization effects, and semiempirical absorption corrections were performed using SADABS [45]. Unit cell parameters at each temperature were determined by the least-squares method with APEX3 software [46]. Structures were solved by direct methods and refined by full-matrix least-squares procedures on F^2 using the WinGX/SHELXT package [47, 48]. Structures, distortion indices, and interatomic angles were illustrated and calculated using

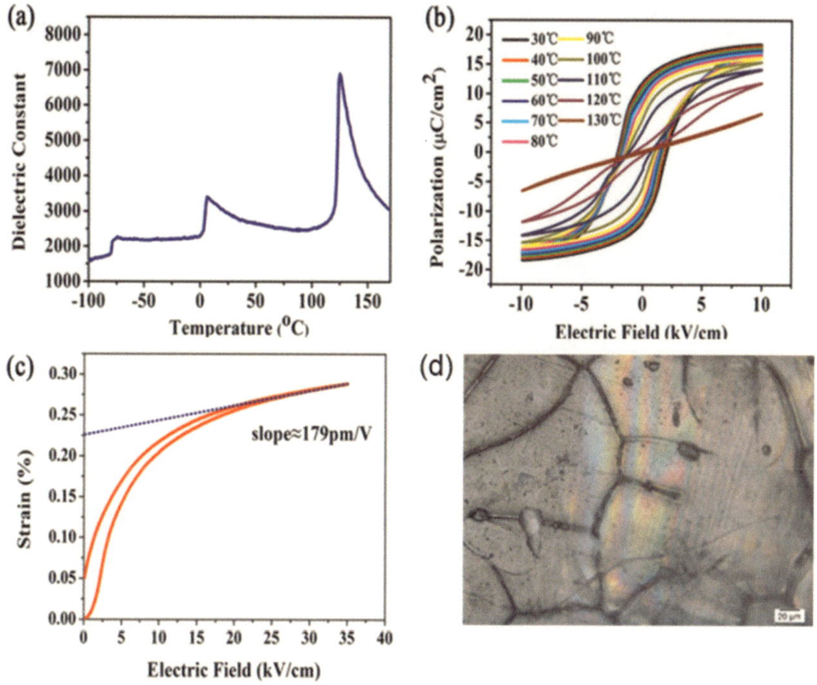

Figure 3.
(a) Dielectric constant as a function of temperature of BCZT single crystal, (b) P-E loops at different temperatures of BCZT single crystal, (c) Unipolar S-E curve of BCZT single crystal, (d) The BCZT single crystal under the plane polarized light microscope. Redrawn from [44], with permission from AIP Publishing.

VESTA3 and Struplo [49, 50]. During refinements, Ba and Ca at the A-site were fixed at the origin position (0, 0, 0) in the non-centrosymmetric space group, and all atoms were refined with anisotropic displacement parameters (U_{ij}).

2.1.1 Temperature-dependent variation in the single crystal structure

The space groups were determined as a function of temperature following Lorentz-polarization and absorption correction. Intensity data indicated point group $3m$ for rhombohedral (−103°C), $mm2$ for orthorhombic (−53°C), and $4mm$ for tetragonal (≥25°C), confirming an intermediate phase as previously reported [36]. Reconstructed precession images of the BCZT single crystal are shown in **Figure 4**, illustrating the space group symmetry in the ($hk0$) plane. Reflection conditions of rhombohedral and tetragonal phases align with previous studies, showing no extinction conditions for $R3m$ and $P4mm$ space groups, respectively [36, 37]. The orthorhombic reflection conditions at 220 K, however, deviate from expected A or B-face-centered ($hkl : k + l = 2n, 0kl : k + l = 2n,$ ... etc.) symmetries, displaying a primitive space group $Pmm2$ [51]. This observation aligns with earlier findings on BTO and BCZT powders, which also reported $Pmm2$ without face centering through X-ray powder diffraction and Raman spectroscopy [42–47, 49–56]. **Figure 4** illustrates these findings, providing a detailed view of the reflection conditions across different phases.

The variation in chemical composition between the single crystal and previously reported powder samples could explain the different space group centering in the orthorhombic phase. Higher Zr content in BCZT50 powder samples directly affects

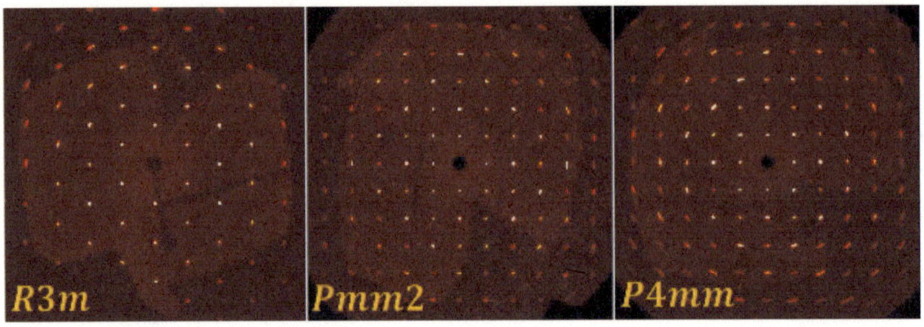

Figure 4.
The reconstructed precession images of BCZT single crystal as a function of symmetry in (hk0) plane. Where R3m, Pmm2, and P4mm space groups were observed at −103, −53, and 25°C. Reprinted from [14], with permission from AIP Publishing.

the A-site local displacement, as evidenced by XAS and EXAFS studies [57, 58]. In our single crystal with low Zr content, Ti ions show displacement due to Zr doping on the B-site, confirmed by XFH studies [16], suggesting Zr ions also displace A-site ions. This might account for the *A*-face-centering super lattice in higher Zr content powders. Similar effects were observed in $Pb(Mg,Nb)O_3$ (PMN) [48], where B-site ion disorder caused A-site Pb ion displacement in the <111> direction. Further studies on high Zr content single crystals are necessary to understand Zr's effect on BCZT symmetry, though growing such crystals is challenging [44, 59].

The lattice parameters and unit cell volume as a function of temperature are detailed in **Tables 3** and **4**. Compared to BTO single crystals in the tetragonal phase at room temperature [62], the *a* axis in BCZT is shorter by 0.0013 Å, and the *c* axis is shorter by 0.0218 Å. The unit cell volume is 0.33 $Å^3$ smaller than BCZT50 powder samples at room temperature [36]. This reduction in lattice parameters and unit cell volume is attributed to different Ca and Zr concentrations. Notably, the slight unit cell volume difference results in significant differences in polarization and the piezoelectric coefficient d_{33}. The c/a ratio in the tetragonal phase decreases with heating, indicating greater a-axis expansion than the c axis. The changes in c/a ratio arise from differential quadratic elongation in the *c* direction and variations in the Ti/ZrO_6 octahedral angles [63]. Quadratic elongation values at 25°C, 47°C, and 107°C are 1.0009, 1.0007, and 0.9995, respectively. Similar effects are observed in BTO and PTO single crystals [62, 64]. The unit cell volume's thermal expansion is $13.00 \times 10^{-4} \pm 1.20 \times 10^{-4}$ $Å^3$/K, lower than the previously reported 18.60×10^{-4} $Å^3$/K for BTO [64]. The thermal expansion contributions are $9.96 \times 10^{-4} \pm 1.70 \times 10^{-4}$ $Å^3$/K from A-site polyhedrons and $2.17 \times 10^{-4} \pm 0.20 \times 10^{-4}$ $Å^3$/K from B-site polyhedrons.

The displacement direction and interatomic angles of the B-site and oxygen ions as a function of symmetry are illustrated in **Figure 5**. In the ac plane, the O–B–O angles show increased octahedral distortion from rhombohedral (89.93°) to orthorhombic (88.30°) upon heating. However, in the tetragonal phase, the angle decreases to 88.70°, nearing cubic phase values. This trend mirrors observations in BTO single crystals [62]. Octahedral rotation, indicated by O–O–O angles in the *ab* plane, also decreases with temperature from rhombohedral (90.46°) to tetragonal (90°). These distortions stem from B-site ion displacement, affecting electron density distribution. The ideal B-site position is (0.5,0.5,0.5) in the cubic phase, shifting to (z,z,z) for rhombohedral and $(0.5,0.5,z)$ for orthorhombic and tetragonal phases, with respective displacements in [111], $[10\bar{1}]$, and $[00\bar{1}]$ directions. The displacement directions

T, °C	−103	−53	25	47	107
Space group	Rhombohedral, *R*3*m**	Orthorhombic, *Pmm*2	Tetragonal, *P*4*mm*	Tetragonal, *P*4*mm*	Tetragonal, *P*4*mm*
Unit cell parameters, Å	$a = 3.9965(4)$, $\alpha = 89.98(1)^{\circ}$	$a = 3.9847(2)$, $b = 4.0056(2)$, $c = 4.0033(2)$	$a = 3.9912(3)$, $c = 4.0153(3)$	$a = 3.9945(1)$, $c = 4.0126(2)$	$a = 4.0020(1)$, $c = 4.0031(2)$
Cell volume V, Å^3	63.83(1)	63.90(1)	63.96(1)	64.03(1)	64.11(1)
Extinction coefficient μ, mm^{-1}	15.57	15.55	15.54	15.52	15.50
F(000)	94.9	94.9	94.9	94.9	94.9
Maximum 2θ, °	72.31	72.46	72.35	72.30	90.53
Data collection range	$-8 \le h \le 9$, $-9 \le k \le 8$, $-11 \le l \le 11$	$-6 \le h \le 6$, $-6 \le k \le 6$, $-6 \le l \le 6$	$-6 \le h \le 6$, $-6 \le k \le 6$, $-6 \le l \le 6$	$-6 \le h \le 6$, $-6 \le k \le 6$, $-6 \le l \le 6$	$-7 \le h \le 8$, $-7 \le k \le 8$, $-8 \le l \le 7$
Total number of reflections	11135	28494	18116	5450	7544
Unique reflections	255	371	209	212	345
R_{int}#, R_1, S	0.054, 0.015, 1.03	0.038, 0.013, 1.16	0.033, 0.012, 1.16	0.048, 0.015, 1.23	0.056, 0.014, 1.23
CSD	2277914	2277915	2277916	2277917	2277918

**R3m space group was refined on hexagonal axes then converted to rhombohedral setting after refinement by using Bilbao Crystallographic Server [60, 61].*
#refers to the merging error (measure of the precision/reproducibility), $R_{int} = \sum |F_0^2 - F_0^2(\text{mean})| / \sum F_0^2$.

Table 3.
Experimental temperature-dependent crystal structure data. Reprinted from [14], with permission from AIP Publishing.

relate to O ion positions, *i.e.*, (x,x,z) for rhombohedral $(0.5,0,z)$, $(0,0.5,z)$, and $(0.5,0.5,z)$; for orthorhombic $(0,0.5,z)$; and for tetragonal $(0.5,0.5,z)$. The deviation from the cubic z value (0.5), reported as Δz, correlates with spontaneous polarization [65, 66]. At 25°C, Δz for B-site ions is 0.008, similar to reported values for BTO (0.010) [67] and BCZT50 powder (0.007) [36] at room temperature.

The change in B-site displacement Δz as a function of temperature is shown in **Figure 6a**. Also presented is an estimate of the z-component of the spontaneous polarization $P_{s,z}$, calculated from the displacements of the B-site and O ions using the equation by Frantti et al. [68]. In the rhombohedral phase, and upon heating, Δz and $P_{s,z}$ increase up to room temperature (tetragonal phase) and then decrease as symmetry approaches the cubic phase. This trend aligns with previously reported properties in BCZT, such as the piezoelectric coefficient d_{33} in bulk samples [21, 69] and remanent polarization in single crystals (**Figure 3**) [44]. However, Keeble et al. reported an opposite trend in BCZT50 [36], with Δz decreasing up to room temperature, and then increasing. In contrast, the T-C transition data for PTO and BTO single crystals showed a similar trend during heating [62, 64]. The differences between single crystal and powder samples could be due to compositional differences, with BCZT50 powder

Atom		−103°C	−53°C	25°C	47°C	107°C
Ba, Ca	x	0.000	0.000	0.000	0.000	0.000
	y	0.000	0.000	0.000	0.000	0.000
	z	0.000	0.000	0.000	0.000	0.000
	U_{eq}	0.006	0.006	0.008	0.008	0.008
Ti, Zr	x	0.497	0.500	0.500	0.500	0.500
	y	0.497	0.500	0.500	0.500	0.500
	z	0.497	0.507	0.492	0.495	0.496
	U_{eq}	0.012	0.012	0.013	0.014	0.013
O1	x	0.496	0.500	0.000	0.000	0.000
	y	0.496	0.000	0.500	0.500	0.500
	z	0.008	0.496	0.503	0.506	0.499
	U_{eq}	0.011	0.012	0.120	0.012	0.011
O2	x	—	0.000	0.500	0.500	0.500
	y	—	0.500	0.500	0.500	0.500
	z	—	0.492	0.013	0.011	−0.004
	U_{eq}	—	0.011	0.012	0.012	0.011
O3	x	—	0.500	—	—	—
	y	—	0.500	—	—	—
	z	—	0.008	—	—	—
	U_{eq}	—	0.011	—	—	—

Table 4.
Atomic positions and thermal coefficient of temperature-dependent crystal structure data.

having higher amounts of substituted elements, like Zr and Ca, compared to BTO and BCZT single crystals. Therefore, further investigation using a model with higher amounts of substituents is essential.

The relative thermal vibrational motions of the ions can be estimated according to the Debye-Waller temperature factor B. As the B-factor is directly related to the mean squared displacement $\langle u^2 \rangle$, the thermal displacement factor U_{eq} of A and B-site ions were reported as a function of temperatures (**Figure 6b**), where $U = \langle u^2 \rangle$ [44]. The U values determined by X-ray diffraction contain both static, *i.e.*, configurational, and dynamic, *i.e.*, thermal vibration, and disorder contributions to the atom according to the Debye model [70]. The anisotropic and equivalent thermal displacement factor U were reported in BCZT single crystal CIF [14]. The configurational disorder can be estimated at 0 K by the intercept of U_{eq} straight line [71, 72]. The calculated intercepts of U_{eq} for A and B-site ions were 0.004(1) Å^2 and 0.0105(8) Å^2, respectively, revealing the existence of configurational disorder in both A- and B-sites in the BCZT single crystal. In contrast, in the case of BTO single crystals, no configurational disorder was observed for Ba and Ti [53]. The differences between BCZT and BTO single crystals are due primarily to the substitution of Ca and Zr on both the A- and B-site, respectively. Thus, the difference in ionic radii between Ca/Ba and Zr/Ti causes this configurational disorder on both sites. Interestingly, the estimated configurational disorder

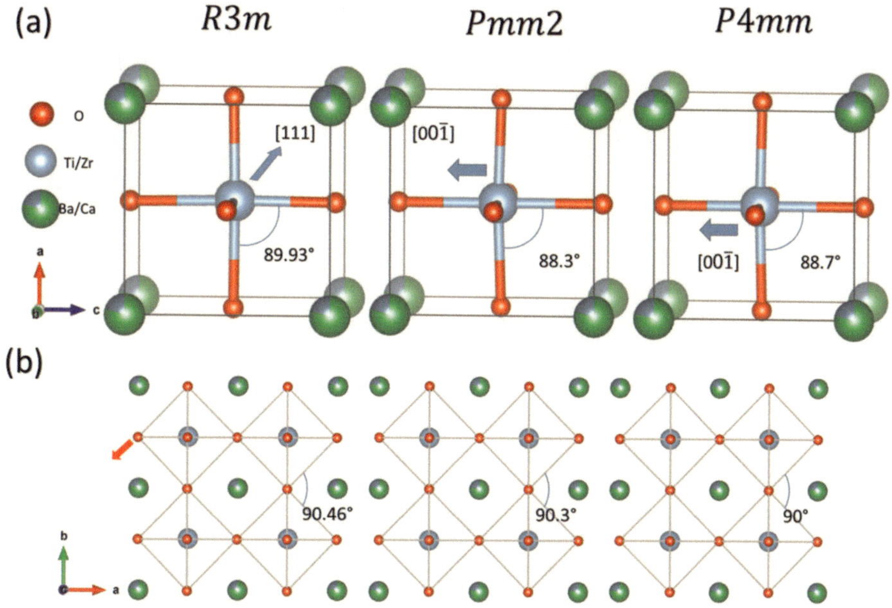

Figure 5.
The crystal structures, the static displacements of the B-site (blue arrows) and O ions (red arrows), and O–B–O and O–O–O interatomic angles as a function of symmetry in (a) ac and (b) ab planes, where R3m, Pmm2, *and* P4mm *were observed at −103, −53, and 25°C. We used VESTA [49] for displaying the crystal structures. Reprinted from [14], with permission from AIP Publishing.*

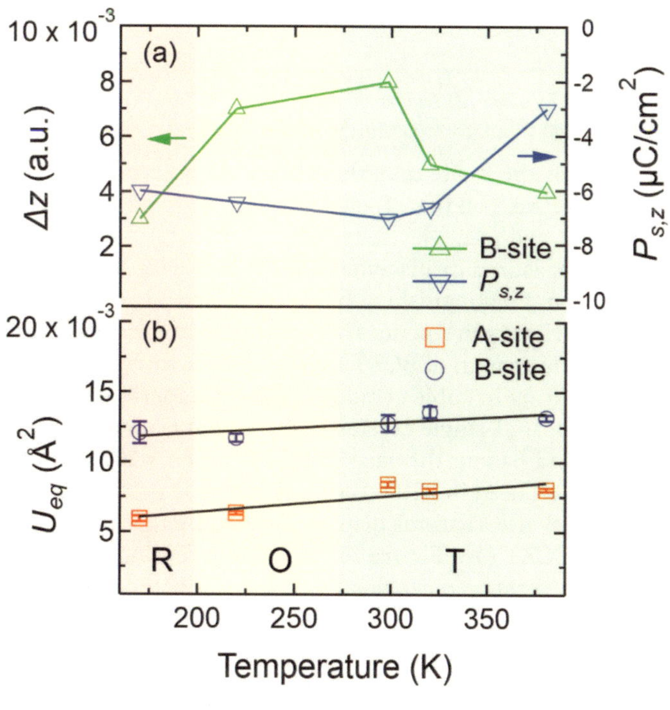

Figure 6.
(a) B-site ion displacement Δz and the spontaneous polarization in z direction and (b) displacement factor U_{eq} *for A and B-site ions as a function of temperature. The phase transitions are based on the published dielectric data from the same sample by Liu* et al. *[44]. Reprinted from [14], with permission from AIP Publishing.*

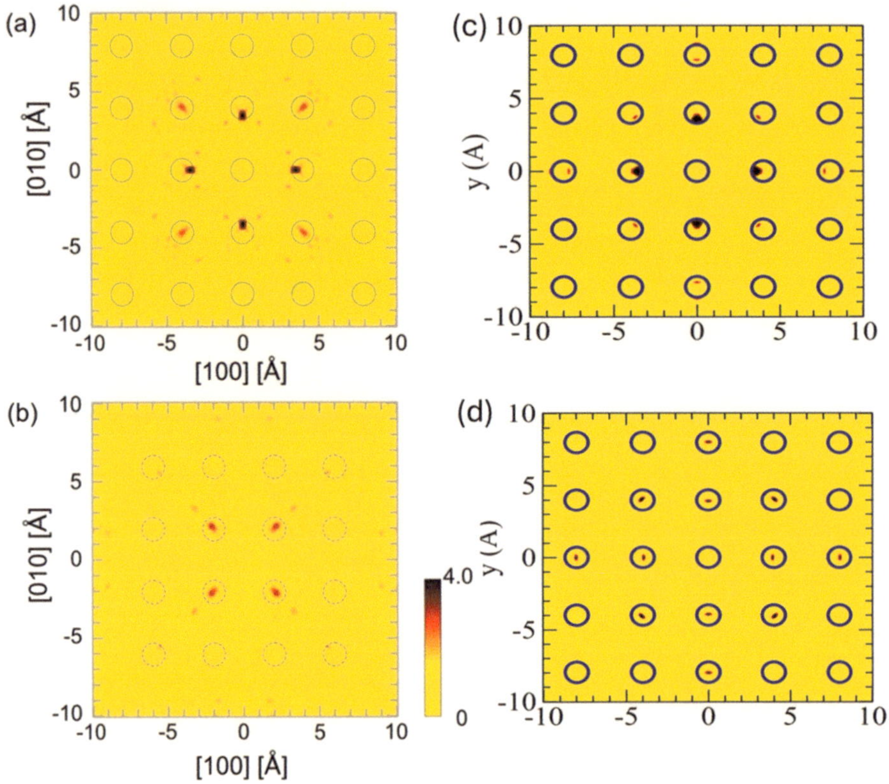

Figure 7.
Atomic images around Zr in BCZT, (a, b) Ti and Ba planes, respectively, as seen from Zr. A-site atomic images on the (001) plane around the (c) Ca and (d) Ba ions obtained from BCT. The circles in the figure show the ideal atomic positions. Reprinted from [16, 17], with permission from Wily and AIP Publishing.

shows a higher value for the B-site than the A-site, as observed in the local displacement by XFH [16, 17]. Based on that, B-site ions are expected to be the main contributor to the polarization and piezoelectric effect.

Interestingly, the measured displacement of the B-site by XRD represents an average displacement, not distinguishing between Zr and Ti at the B-site or Ca and Ba at the A-site. Thus, XFH measurements are essential to understand the effects of Ca and Zr in both sites. The growth of BCZT single crystals with high Zr content is challenging, often resulting in cubic crystals at room temperature [39–43]. Inverse mode measurements on BCT single crystals showed that Ca ions displace by 0.36 Å in the <111> direction, and Ba ions displace by 0.22 Å in the <100> direction [14]. Low Zr content BCZT crystals used in XRD studies reveal that A-site atoms around Zr displace by 0.2 Å, and B-site Ti atoms displace inward by 0.5 Å, with no such displacement observed in CZT (see **Figure 7**) [16]. These findings are crucial for understanding substitution effects on piezoelectric properties compared to $BaTiO_3$.

3. Conclusion

In this chapter, the crystal structure and symmetry of $BaTiO_3$-based ferroelectric material was reported for single BCZT. Beside the low-temperature rhombohedral and room temperature tetragonal phases, the existence of an intermediate orthorhombic phase in BCZT was confirmed. The displacement of B-site cations correlates directly with spontaneous polarization, and configurational disorder was observed for both A- and B-site cations. The displacement factor U_{eq} of A and B-site cations indicates both configurational and thermal vibration disorders due to the substitutions. These distortions explain changes in symmetry as a function of composition. These studies provide crucial insights into the role of Ba/Ca and Ti/Zr substitutions at the A- and B-sites in $BaTiO_3$, respectively, to understand their direct effects on the crystal structure and phase transitions of BCZT.

Author details

Ahmed Gadelmawla
Department of Materials Science and Engineering, Friedrich-Alexander-Universität Erlangen-Nürnberg (FAU), Erlangen, Germany

*Address all correspondence to: ahmed.gadelmawla@fau.de

References

[1] Uchino K. Ferroelectric Devices. 2nd ed. Hoboken: Taylor and Francis; 2009. Available from: http://gbv.eblib.com/patron/FullRecord.aspx?p=1388860

[2] Carter CB, Norton MG. Ceramic Materials: Science and Engineering. New York, NY, USA: Springer; 2007

[3] Moulson AJ, Herbert JM. Electroceramics: Materials, Properties, Applications. New York: Wiley; 2003. Available from: http://onlinelibrary.wiley.com/book/10.1002/0470867965

[4] Cook WR, Jaffe H, Jaffe B. Piezoelectric Ceramics. New York, NY, USA: Academic Press; 1971

[5] Webber KG, Vögler M, Khansur NH, Kaeswurm B, Daniels JE, Schader FH. Review of the mechanical and fracture behavior of perovskite lead-free ferroelectrics for actuator applications. Smart Materials and Structures. 2017;**26**(6):63001. DOI: 10.1088/1361-665X/aa590c

[6] Rödel J, Webber KG, Dittmer R, Jo W, Kimura M, Damjanovic D. Transferring lead-free piezoelectric ceramics into application. Journal of the European Ceramic Society. 2015;**35**(6):1659-1681. DOI: 10.1016/j.jeurceramsoc.2014.12.013

[7] Gadelmawla AS. Stress modulated electromechanical coupling of lead-free ferroelectrics [doctoral dissertation]. Erlangen: Friedrich-Alexander-Universität Erlangen-Nürnberg (FAU); 2023

[8] Neumann FE. Vorlesungen über die Theorie der Elastizität der festen Körper und des Lichtäthers. Leipzig: B. G. Teubner; 1885

[9] Hahn T. International Tables for Crystallography. 5th ed. Dordrecht: Springer; 2005. Available from: http://deposit.dnb.de/cgi-bin/dokserv?id=2730166&prov=M&dok_var=1&dok_ext=htm

[10] Nye JF. Physical Properties of Crystals: Their Representation by Tensors and Matrices. Oxford: Oxford University Press; 1985

[11] Mitzi DB. Synthesis, structure, and properties of organic—Inorganic perovskites and related materials. ChemInform. 1999;**30**(31):1-34

[12] Goldschmidt R. The Material Basis of Evolution. Vol. 28. New Haven: Yale University Press; 1982

[13] Liu W, Ren X. Large piezoelectric effect in Pb-free ceramics. Physical Review Letters. 2009;**103**(25):257602. DOI: 10.1103/PhysRevLett.103.257602

[14] Gadelmawla A et al. Investigation of ferroelectric $Ba_{1-x}Ca_xZr_yTi_{1-y}O_3$ single crystal by in situ temperature-dependent x-ray diffraction and first-principles calculations. Journal of Applied Physics. 2023;**134**(14):144101-144114

[15] Dobesh DK et al. The role of Ca/Zr ratio on the local structure and phase transitions in lead-free (Ba, Ca)(Zr, Ti)O_3. Journal of the European Ceramic Society. 2024;**44**(10):5646-5658

[16] Yamamoto Y et al. Local structural investigation of (Ba,Ca)(Zr,Ti)O_3 and Ca(Zr,Ti)O_3 by X-ray fluorescence holography. Physica Status Solidi B: Basic Solid State Physics. 2022;**259**(9):2100609. DOI: 10.1002/pssb.202100609

[17] Yamamoto Y et al. Significant displacement of calcium and barium ions

in ferroelectric $(Ba_{0.9}Ca_{0.1})TiO_3$ revealed by x-ray fluorescence holography. Applied Physics Letters. 2022;**120**(5):52905. DOI: 10.1063/5.0076325

[18] Bai Y et al. (Ba,Ca)(Zr,Ti)O_3 lead-free piezoelectric ceramics—The critical role of processing on properties. Journal of the European Ceramic Society. 2015; **35**(13):3445-3456. DOI: 10.1016/j.jeurceramsoc.2015.05.010

[19] Kay HF. An Investigation of the Structure and Properties of the Titanates of Barium, Strontium and Calcium. Available from: https://search.proquest.com/openview/c11c08c80e46c9d79d89208bb29996e7/1?pq-origsite=gscholar&cbl=2026366&diss=y

[20] Devonshire AF. Theory of barium titanate. The London, Edinburgh, and Dublin Philosophical Magazine Journal of Science. 1949;**40**(309):1040-1063. DOI: 10.1080/14786444908561372

[21] Acosta M et al. Origin of the large piezoelectric activity in $(1-x)Ba(Zr_{0.2}Ti_{0.8})O_3-x(Ba_{0.7}Ca_{0.3})TiO_3$ ceramics. Physical Review B. 2015;**91**(10):104108-104117. DOI: 10.1103/PhysRevB.91.104108

[22] Zhang R, Li JF, Viehland D. Effect of aliovalent substituents on the ferroelectric properties of modified barium titanate ceramics—Relaxor ferroelectric behavior. Journal of the American Ceramic Society. 2004;**87**(5):864-870. DOI: 10.1111/j.1551-2916.2004.00864.x

[23] Darlington CNW, Cernik RJ. The effects of isovalent and non-isovalent impurities on the ferroelectric phase transition in barium titanate. Journal of Physics. Condensed Matter. 1993;**5**(32):5963-5970. DOI: 10.1088/0953-8984/5/32/024

[24] Rahaman MN, Manalert R. Grain boundary mobility of $BaTiO_3$ doped with aliovalent cations. Journal of the European Ceramic Society. 1998;**18**(8):1063-1071. DOI: 10.1016/S0955-2219(97)00215-X

[25] Schader FH, Aulbach E, Webber KG, Rossetti GA. Influence of uniaxial stress on the ferroelectric-to-paraelectric phase change in barium titanate. Journal of Applied Physics. 2013;**113**(17):174103. DOI: 10.1063/1.4799581

[26] Liu QJ, Zhang NC, Liu FS, Wang HY, Liu ZT. $BaTiO_3$: Energy, geometrical and electronic structure, relationship between optical constant and density from first-principles calculations. Optical Materials. 2013;**35**(12):2629-2637. DOI: 10.1016/j.optmat.2013.07.034

[27] Jia X, Zhang HQ, Wang Z, Jiang CL, Liu QJ, Liu ZT. Mechanical properties, born effective charge tensors and high frequency dielectric constants of the eight phases of $BaTiO_3$. Moscow University Physics. 2017;**72**(4):358-363. DOI: 10.3103/S0027134917040154

[28] Park JG, Oh TS, Kim YH. Dielectric properties and microstructural behaviour of B-site calcium-doped barium titanate ceramics. Journal of Materials Science. 1992;**27**(21):5713-5719. DOI: 10.1007/BF01119727

[29] Ganguly M et al. Characterization and Rietveld refinement of A-site deficient lanthanum doped barium titanate. Journal of Alloys and Compounds. 2013;**579**:473-484. DOI: 10.1016/j.jallcom.2013.06.104

[30] Verbitskaya TN, Svetlova LN, Sokolova LS, Laverko EN, Raevskaya EB. Influence of added calcium titanate on the structure and properties of ferroelectric solid solutions based on

$BaTiO_3$. Soviet Physics Crystallography. 1988;**33**:609-610

[31] Kumar P, Singh S, Juneja JK, Prakash C, Raina KK. Influence of calcium substitution on structural and electrical properties of substituted barium titanate. Ceramics International. 2011;**37** (5):1697-1700

[32] Varatharajan R, Samanta S, Jayavel R, Subramanian C, Narlikar A, Ramasamy P. Ferroelectric characterization studies on barium calcium titanate single crystals. Materials Characterization. 2000;**45**(2):89-93. DOI: 10.1016/S1044-5803(00)00053-X

[33] Tang XG, Chew KH, Chan H. Diffuse phase transition and dielectric tunability of Ba(ZryTi1−y)O_3 relaxor ferroelectric ceramics. Acta Materialia. 2004;**52**(17):5177-5183. DOI: 10.1016/j.actamat.2004.07.028

[34] Veerapandiyan V et al. Origin of Relaxor behavior in barium-titanate-based Lead-free perovskites. Advanced Electronic Materials. 2022;**8**(2):2100812. DOI: 10.1002/aelm.202100812

[35] Shannon RD. Revised effective ionic radii and systematic studies of interatomic distances in halides and chalcogenides. Acta Crystallographica A. 1976;**32**(5):751-767. DOI: 10.1107/S0567739476001551

[36] Keeble DS, Benabdallah F, Thomas PA, Maglione M, Kreisel J. Revised structural phase diagram of $(Ba_{0.7}Ca_{0.3}TiO_3)$-$(BaZr_{0.2}Ti_{0.8}O_3)$. Applied Physics Letters. 2013;**102**(9): 92903. DOI: 10.1063/1.4793400

[37] Bjørnetun Haugen A, Forrester JS, Damjanovic D, Li B, Bowman KJ, Jones JL. Structure and phase transitions in 0.5 $(Ba_{0.7}Ca_{0.3}TiO_3)$-0.5$(BaZr_{0.2}Ti_{0.8}O_3)$ from −100°C to 150 °C. Journal of Applied Physics. 2013;**113**(1):14103. DOI: 10.1063/1.4772741

[38] Acosta M et al. $BaTiO_3$-based piezoelectrics: Fundamentals, current status, and perspectives. Applied Physics Reviews. 2017;**4**(4):41305. DOI: 10.1063/1.4990046

[39] Zeng Y, Zheng Y, Tu X, Lu Z, Shi E. Growth and characterization of lead-free $Ba_{(1-x)}Ca_xTi_{(1-y)}Zr_yO_3$ single crystal. Journal of Crystal Growth. 2012;**343**(1): 17-20. DOI: 10.1016/j.jcrysgro.2012.01.009

[40] Bhaumik I et al. Growth of lead-free piezoelectric 0.45BZT–0.55BCT single crystal and investigation of dielectric, polarization and birefringence properties. Journal of Crystal Growth. 2013;**375**:20-25. DOI: 10.1016/j.jcrysgro.2013.04.033

[41] Singh G et al. Electro-caloric effect in $0.45BaZr_{0.2}Ti_{0.8}O_3$-$0.55Ba_{0.7}Ca_{0.3}TiO_3$ single crystal. Applied Physics Letters. 2013;**102**(8):82902. DOI: 10.1063/1.4793213

[42] Benabdallah F, Veber P, Prakasam M, Viraphong O, Shimamura K, Maglione M. Continuous cross-over from ferroelectric to relaxor state and piezoelectric properties of $BaTiO_3$-$BaZrO_3$-$CaTiO_3$ single crystals. Journal of Applied Physics. 2014;**115**(14): 144102. DOI: 10.1063/1.4870933

[43] Palneedi H et al. Strong and anisotropic magnetoelectricity in composites of magnetostrictive Ni and solid-state grown lead-free piezoelectric BZT–BCT single crystals. Journal of Asian Ceramic Societies. 2017;**5**(1):36-41. DOI: 10.1016/j.jascer.2016.12.005

[44] Liu D, Shim J, Sun Y, Li Q, Yan Q. Growth of Ca, Zr co-doped $BaTiO_3$ lead-free ferroelectric single crystal and its

room-temperature piezoelectricity. AIP Advances. 2017;**7**(9):95311. DOI: 10.1063/1.4994764

[45] Krause L, Herbst-Irmer R, Sheldrick GM, Stalke D. Comparison of silver and molybdenum microfocus X-ray sources for single-crystal structure determination. Journal of Applied Crystallography. 2015;**48**:3-10. DOI: 10.1107/S1600576714022209

[46] Bruker AXS, Inc., 2018, Madison, WI, U.S.A

[47] Farrugia LJ. WinGX and ORTEP for windows: An update. Journal of Applied Crystallography. 2012;**45**(4):849-854. DOI: 10.1107/S0021889812029111

[48] Hu W et al. Acute and obtuse rhombohedrons in the local structures of relaxor ferroelectric $Pb(Mg1/3Nb2/3)O_3$. Physical Review B. 2014;**89**(14):140103. DOI: 10.1103/PhysRevB.89.140103

[49] Momma K, Izumi F. VESTA 3 for three-dimensional visualization of crystal, volumetric, and morphology data. Journal of Applied Crystallography. 2011;**44**(6):1272-1276. DOI: 10.1107/S0021889811038970

[50] Fischer RX. STRUPLO 84, a Fortran plot program for crystal structure illustrations in polyhedral representation. Journal of Applied Crystallography. 1985;**18**(4):258-262. DOI: 10.1107/S002188988501024X

[51] Buerger MJ. The application of plane groups to the interpretation of Weissenberg photographs. Zeitschrift Kristallographie Crystalline Materials. 1935;**91**:255-289. DOI: 10.1524/zkri.1935.91.1.255

[52] Sheldrick GM. SHELXT—Integrated space-group and crystal-structure determination. Acta Crystallographica Section A. 2015;**71**(1):3-8. DOI: 10.1107/S2053273314026370

[53] Abdessalem MB et al. Polymorphic phase transition and morphotropic phase boundary in $Ba_{1-x}\,Ca_x\,Ti_{1-y}\,Zr_y\,O_3$ ceramics. Applied Physics A: Materials Science & Processing. 2017;**123**(9):686-696. DOI: 10.1007/s00339-017-1196-7

[54] Kaddoussi H et al. Sequence of structural transitions and electrocaloric properties in $(Ba_{1-x}Ca_x)(Zr_{0.1}Ti_{0.9})O_3$ ceramics. Journal of Alloys and Compounds. 2017;**713**:164-179. DOI: 10.1016/j.jallcom.2017.04.148

[55] Nayak RL, Zhang Y, Dash SS, Sahoo M. Intermediate ferroelectric phase driven ferroelectric-relaxor crossover and superior storage properties in phase boundary engineered BCZT ceramics. Ceramics International. 2022;**48**(8):10803-10816. DOI: 10.1016/j.ceramint.2021.12.296

[56] Zeng L, Jiang Y. X-ray diffraction study of orthorhombic barium titanate. Beijing Gongye Daxue Xuebao. 1991;**17**:32-37

[57] Manjón-Sanz A et al. Total scattering and diffraction studies of lead-free piezoelectric $(1\text{-}x)Ba(Zr_{0.2}Ti_{0.8})O_{3\text{-}x}(Ba_{0.7}Ca_{0.3})TiO_3$ deconvolute intrinsic and extrinsic contributions to electromechanical strain. Acta Materialia. 2019;**171**:79-91. DOI: 10.1016/j.actamat.2019.04.005

[58] Munthala D et al. In-situ X-ray absorption spectroscopy (XAS) studies of electrical field induced domain switching in BCZT ceramics. Ceramics International. 2021;**47**(17):25158-25162. DOI: 10.1016/j.ceramint.2021.05.201

[59] Sun Y et al. Piezoelectric property of a tetragonal $(Ba,Ca)(Zr,Ti)O_3$ single crystal and its fine-domain structure.

ACS Applied Materials & Interfaces. 2018;**10**(15):12847-12853. DOI: 10.1021/acsami.8b01360

[60] Aroyo MI et al. Bilbao crystallographic server: I. Databases and crystallographic computing programs. Zeitschrift Kristallographie Crystalline Materials. 2006;**221**(1):15-27. DOI: 10.1524/zkri.2006.221.1.15

[61] Aroyo MI et al. Representations of crystallographic point groups and space groups. Acta Crystallographica Section A. 2006;**62**(Pt 2):115-128. DOI: 10.1107/S0108767305040286

[62] Nakatani T et al. Variable-temperature single-crystal X-ray diffraction study of tetragonal and cubic perovskite-type barium titanate phases. Acta Crystallographica. Section B. 2016; **72**(Pt 1):151-159. DOI: 10.1107/S2052520615022544

[63] Robinson K, Gibbs GV, Ribbe PH. Quadratic elongation: A quantitative measure of distortion in coordination polyhedra. Science. 1971;**172**(3983):567-570. DOI: 10.1126/science.172.3983.567

[64] Yoshiasa A et al. High-temperature single-crystal X-ray diffraction study of tetragonal and cubic perovskite-type $PbTiO_3$ phases. Acta Crystallographica. Section B. 2016;**72**(Pt 3):381-388. DOI: 10.1107/S2052520616005114

[65] Evans HT. The crystal structure of tetragonal barium titanate. Acta Cryst. 1951;**4**(4):377. DOI: 10.1107/S0365110X51001173

[66] Evans HT. An X-ray diffraction study of tetragonal barium titanate. Acta Cryst. 1961;**14**(10):1019-1026. DOI: 10.1107/S0365110X61002989

[67] Buttner RH, Maslen EN. Electron difference density and vibration tensors in $SrTiO_3$. Acta Crystallographica. Section B. 1992;**48**(5):639-644. DOI: 10.1107/S0108768192004488

[68] Frantti J et al. Phase transitions of Pb$(Zr_xTi_{1-x})O_3$ ceramics. Physical Review B. 2002;**66**(6):64108. DOI: 10.1103/PhysRevB.66.064108

[69] Gadelmawla A et al. Influence of stress on the electromechanical properties and the phase transitions of lead-free $(1-x)Ba(Zr_{0.2}Ti_{0.8})O_3$–x$(Ba_{0.7}Ca_{0.3})TiO_3$. Journal of Materials Science. 2022;**57**:1-19. DOI: 10.1007/s10853-022-07685-9

[70] Willis BTM, Pryor AW. Thermal Vibrations in Crystallography. Cambridge, England: Cambridge University Press; 1975

[71] Yoshiasa A, Koto K, Kanamaru F, Emura S, Horiuchi H. Anharmonic thermal vibrations in wurtzite-type AgI. Acta Crystallographica. Section B. 1987; **43**(5):434-440. DOI: 10.1107/S0108768187097532

[72] Nakatsuka A et al. Static disorders of ions and experimental determination of debye temperature in pyrope: Low- and high-temperature single-crystal X-ray diffraction study. American Mineralogist. 2011;**96**(10):1593-1605. DOI: 10.2138/am.2011.3714

Chapter 4

Polarization and Ferromagnetism in Microwave-Absorbing Materials

Udeshwari Jamwal, Shivam Kumar Mittal and Deepanshu Keneria

Abstract

This chapter explores the intricate relationship between electric polarization and ferromagnetic properties in microwave-absorbing materials. It highlights the fundamental principles of dielectric and magnetic losses that contribute to microwave absorption. Emphasizing composite materials that exhibit both ferroelectric and ferromagnetic characteristics, it delves into the mechanisms of electric polarization and how they interact with magnetic domains to enhance microwave absorption. By examining the coupling between ferroelectric and magnetic properties, the chapter aims to uncover the synergistic effects that optimize the performance of microwave absorbers. Key topics include the synthesis and characterization of these materials and the role of different loss mechanisms in absorbing materials. The applications of these materials in stealth technology, electromagnetic interference shielding, and wireless communication are discussed, showcasing their significance in modern technological advancements.

Keywords: electric polarization, ferromagnetic properties, microwave absorption, magnetic domain, electric polarization mechanisms, ferroelectric-magnetic coupling, composite materials

1. Introduction

The rapid advancement of modern technology has led to an increased need for efficient microwave-absorbing materials (MAMs) [1]. These materials are crucial in a wide range of applications, from reducing electromagnetic interference (EMI) in electronic devices to enhancing the stealth capabilities of military equipment [2]. MAMs are designed to attenuate electromagnetic waves, thereby preventing reflection or transmission of these waves. This capability is essential in many sectors, including telecommunications, automotive, aerospace, and defense industries [3]. The study of MAMs is not new; it has evolved significantly over the past few decades. Initially, research focused on simple materials and their basic properties. However, with the advent of new technologies and the increasing complexity of modern systems, there has been a shift toward more advanced materials that can offer superior performance. This shift has brought about the need to understand and manipulate the underlying mechanisms that contribute to microwave absorption [4–6]. MAMs work on the principle of converting electromagnetic energy into heat, which is then dissipated [7]. This process relies on two main mechanisms: dielectric losses and magnetic losses. Dielectric losses occur from the polarization of electric dipoles, while magnetic losses arise from the alignment and movement of magnetic domains. Efficient MAMs balance these losses, with high dielectric loss materials absorbing electric fields and high magnetic loss materials absorbing magnetic fields. The challenge is developing materials that exhibit both high dielectric and magnetic losses to maximize microwave absorption. Composite materials combining ferroelectric and ferromagnetic properties offer a solution by leveraging both types of losses. Fine-tuning these composites can optimize microwave absorption across a wide frequency range. Electric polarization and ferromagnetic properties significantly influence MAM performance. Electric polarization, the alignment of electric dipoles in response to an external electric field, creates polarization currents that contribute to dielectric losses. Ferromagnetic properties, involving the alignment of magnetic domains, result in magnetic losses essential for absorbing magnetic components of electromagnetic waves. Understanding and manipulating these properties enables the design of materials with tailored absorption characteristics, improving microwave absorption efficiency. This knowledge also aids in developing new materials and technologies, with broader implications for fields like energy storage, sensors, and actuators. Advances in these areas can lead to innovations across various technological domains.

This chapter aims to provide a comprehensive exploration of the relationship between electric polarization and ferromagnetic properties in MAMs. It begins by detailing the principles of microwave absorption, focusing on dielectric and magnetic losses, and examines the benefits of combining ferroelectric and ferromagnetic materials in composites. Synthesis and characterization methods are discussed, along with the challenges in optimizing microwave absorption. The core focus is on how electric polarization and magnetic domains interact to enhance absorption efficiency. Practical applications in stealth technology, EMI shielding, and wireless communication are highlighted. The chapter concludes with future trends, emerging materials, challenges, and research opportunities, providing a thorough understanding of microwave absorption mechanisms and advancements.

2. Fundamentals of electric polarization and ferromagnetism

2.1 Electric polarization

Electric polarization occurs when electric dipoles within a material align in response to an external electric field, resulting in a net dipole moment per unit volume [8]. This phenomenon can be classified into various types based on the alignment mechanism and cause of polarization. Spontaneous polarization arises in materials with an inherent permanent dipole moment, even without an external electric field, as seen in ferroelectric materials like barium titanate ($BaTiO_3$), which have a non-centrosymmetric arrangement of ions. Induced polarization, on the other hand, occurs when an external electric field generates a dipole moment in the material. This can be further divided into electronic polarization, ionic polarization, orientation polarization, and space-charge polarization, each describing different mechanisms of dipole alignment in response to the applied field [9].

2.1.1 Mechanisms of electric polarization

The mechanisms by which electric polarization occurs in materials are diverse and depend on the material's structure and the nature of the external electric field. The total polarization of a dielectric (P) comes from the sum of four sources of polarization: electronic polarization P_e, ionic polarization P_i, orientation of permanent dipoles P_o, and space-charge polarization P_{sc}

$$P = P_e + P_i + P_o + P_{sc} \tag{1}$$

These four polarization mechanisms are described as follows:

- *Electronic polarization*: Electronic polarization occurs when an external electric field displaces the negatively charged electron cloud relative to the positively charged nucleus, generating a dipole moment. This high-frequency polarization involves the distortion of the electron cloud around atoms or molecules and is the fastest and most reversible mechanism, typically occurring on the timescale of electronic transitions [10]. The mechanism of electronic polarization is shown in **Figure 1(a)**.

The induced dipole moment due to an applied electric field (E) on the atom is $p = \alpha E$, where the constant of proportionality α is called atomic polarizability.

The polarization (P), defined as the total dipole moment per unit volume, is:

$$P = Np = N\alpha E = \chi_e \varepsilon_o E \tag{2}$$

where N is the number of molecules per unit volume, χ_e is electric susceptibility, ε_0 is the permittivity of free space, and E is the applied electric field. Using the equations.

$\chi_e = \varepsilon_r - 1$ and $E_{loc} = E + \dfrac{p}{3\varepsilon 0}$, the above equation leads to the Clausius-Mossotti equation [11]:

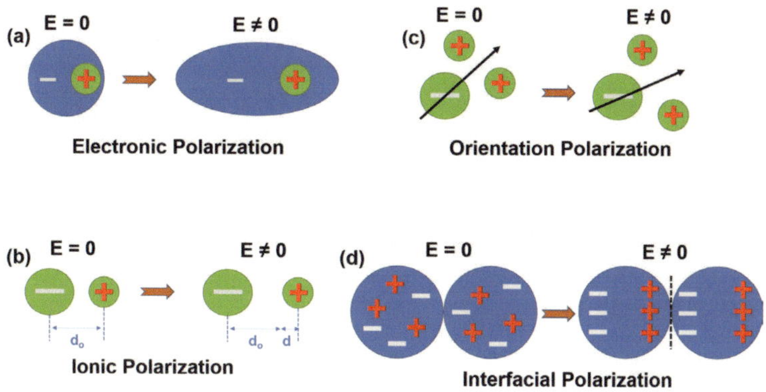

Figure 1.
Schematic Diagrams of (a) electronic polarization, (b) ionic polarization, (c) orientation polarization, and (d) interfacial polarization.

$$\frac{\varepsilon_r - 1}{\varepsilon_r + 2} = \frac{N\alpha_e}{3\varepsilon_0} \tag{3}$$

where ε_r is relative permittivity, and E_{loc} is the local electric field.

This is the Clausius-Mossotti equation, which relates the macroscopic dielectric constant (relative permittivity) of a material to the microscopic polarizability of its constituent molecules or atoms.

Frequency range: Typically, from optical frequencies (infrared region) to ultraviolet (UV) region, typically up to $10^{15} - 10^{17}$ Hz.

- *Ionic polarization*: Ionic polarization occurs in ionic crystals like NaCl, KCl, and LiBr. Without an external electric field, the dipole moments of negative and positive ions cancel each other out, resulting in no net polarization. When an external field is applied, positive and negative ions shift from their equilibrium positions, creating a dipole moment. This displacement is opposed by the lattice's restoring force, resulting in net polarization. This process is slower than electronic polarization and significant at lower frequencies. The equation to describe this effect is given by,

$$p_{av} = \alpha_i E_{loc} \tag{4}$$

where p_{av} is the induced average dipole moment per ion pair, α_i is the ionic polarizability. Usually, the ionic polarizability is greater than the electronic polarizability by a factor of 10 which leads to ionic substances having high dielectric constants. Similar to electronic polarization, ionic polarization also has a total polarization associated with it. The equation is given by

$$P_i = Np_{av} = N\alpha_i E_{loc} \tag{5}$$

Frequency range: Typically, up to the infrared region, around 10^{12}–10^{14} Hz.

- *Orientation polarization*: In materials containing polar molecules, an external electric field aligns the dipoles in the direction of the field. The degree of alignment depends on the thermal energy of the system and the strength of the electric field. This process is slower than electronic and ionic polarization and can exhibit relaxation effects when the field is removed. It is most active in the radio to microwave frequency ranges. This type is frequency-dependent and can relax when the field is removed.

The Polarization is given by

$$P = N = Np_o \tag{6}$$

in this case, the average dipole moment is given by

$$p_{av} = \frac{p_0^2 E}{KT} = \alpha_d E \tag{7}$$

where p_0 is the dipole moment of a dipole that gets oriented in the direction of field, K is the Boltzmann constant, T is temperature α_d is the dipolar polarizability.

Frequency range: Typically, from microwave frequencies down to a few kilohertz, around 10^3–10^{10} Hz.

- *Interfacial (space charge) polarization*: This occurs in heterogeneous materials or those with defects, where the migration of charges leads to the accumulation at interfaces or defects, creating a polarization effect, as shown in **Figure 1(d)**. This type is particularly important in materials with grain boundaries, dislocations, or other imperfections [10].

Frequency Range: It occurs from the frequency range from low frequencies (a few hertz) to around 103–105 Hz.

- *Dielectric constant v/s frequency*: At lower frequencies, the dielectric constant is notably high, likely due to the accumulation of interfacial charge carriers and the contributions of all polarizations (such as dipolar, ionic and electronic), as shown in **Figure 2**. As the frequency increases, the dielectric constant decreases. This can be attributed to the electric dipole's inability to respond to the applied field and the fact that charge carriers cannot quickly accumulate at the interface, leaving only electronic polarization to contribute [12].

2.2 Polarization in dielectric and ferroelectric materials

Dielectric materials are insulators that can be polarized under an electric field, resulting in stored electric energy. The dielectric constant, a measure of a material's ability to be polarized, is a key property of these materials. Dielectrics can be linear or non-linear, depending on the relationship between the polarization and the applied electric field. Ferroelectric materials are a subset of dielectrics that exhibit spontaneous polarization that can be reversed by an external electric field. These materials have a characteristic hysteresis loop, representing the relationship between polarization and electric field [13]. The presence of a permanent dipole moment even in the

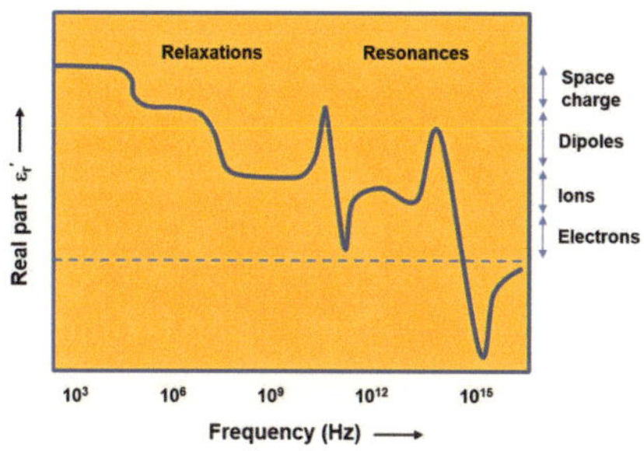

Figure 2.
Frequency-dependent polarization mechanisms.

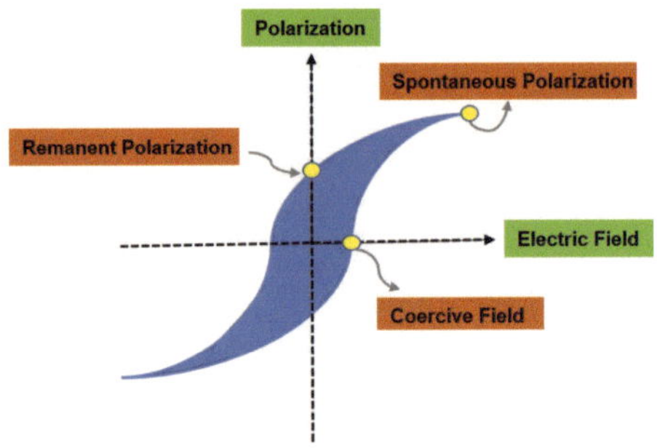

Figure 3.
Typical polarization vs. electric field (P-E) hysteresis loop of ferroelectrics.

absence of an external field known as spontaneous polarization, is shown in **Figure 3**. The dependence of polarization on the history of the applied electric field results in a loop when plotted, which is called hysteresis. Ferroelectric materials are widely used in applications such as non-volatile memory, capacitors, and piezoelectric devices due to their unique polarization properties.

2.3 Ferromagnetism

Ferromagnetism is a phenomenon where materials retain permanent magnetic moments even in the absence of an external magnetic field due to the alignment of magnetic moments within specific regions known as magnetic domains. In these materials, magnetic moments—vector quantities representing the strength and orientation of the magnetism—align parallel to each other within domains, separated by boundaries called domain walls. The Curie temperature is the point at which a ferromagnetic material loses its permanent magnetism and transitions to a paramagnetic state, as thermal energy disrupts the alignment of these moments. The process of magnetization involves aligning the domains with an external magnetic field, and the extent of this alignment reflects the material's responsiveness to the field.

- *Origin of ferromagnetism in materials*: The origin of ferromagnetism lies in the quantum mechanical exchange interactions between neighboring atoms or ions. These interactions are a result of the Pauli exclusion principle and the Coulomb repulsion between electrons [14]. The key mechanisms that contribute to ferromagnetism include:

- *Exchange interaction*: The exchange interaction, shown in **Figure 4(a)**, is a quantum mechanical effect that arises from the wavefunction overlap of neighboring electrons. It arises due to the Pauli exclusion principle and the Coulomb interaction between electrons, leading to a preference for certain spin alignments that minimize the system's energy. This interaction is fundamental to understanding various types of magnetism in materials, including ferromagnetism, antiferromagnetism, and ferrimagnetism.

There are several categories of exchange interactions, each with distinct mechanisms and implications for magnetic behavior.

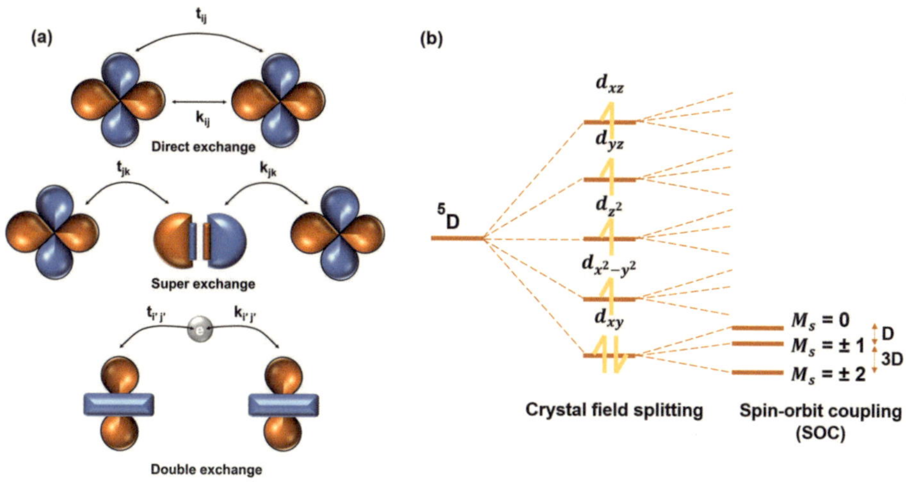

Figure 4.
Schematic illustration of (a) Exchange Interaction, and (b) Spin-orbit coupling, and Crystal field Effects.

a. *Direct exchange*: This interaction occurs when the wavefunctions of electrons on adjacent atoms overlap, allowing the electrons to "sense" each other's spin and create a magnetic interaction. The strength and alignment (parallel or antiparallel) of this interaction depend on the distance between atoms and the symmetry of their wavefunctions. If the interaction energy is minimized with parallel spins, the material exhibits ferromagnetism. Parallel spins lower the system's energy due to favorable electron overlap and reduced Coulomb repulsion. If the interaction energy is minimized with antiparallel spins, the material exhibits antiferromagnetism. Antiparallel spin alignment lowers the system's energy, often due to specific spatial arrangements of atoms that favor this configuration.

b. *Superexchange*: It is an indirect exchange interaction occurring in insulating materials where magnetic ions are separated by a non-magnetic anion, such as oxygen. In this mechanism, the interaction between the magnetic moments of the ions is mediated by the non-magnetic anion through a process known as virtual electron hopping. This process effectively creates a magnetic interaction between the ions, even though they are not directly adjacent. Superexchange often results in antiferromagnetic interactions, particularly when the bond angle between the magnetic ions is close to 180°. This bond angle facilitates effective electron hopping via the anion, leading to an antiparallel alignment of magnetic moments and minimizing the system's energy.

c. *Double exchange*: This mechanism that occurs in mixed-valence compounds where two magnetic ions of the same element exist in different oxidation states (e.g., Mn^{3+} and Mn^{4+}) The interaction is mediated by the transfer of an electron between the ions, and this transfer is facilitated when the spins are aligned, leading to ferromagnetic ordering. In double exchange, the alignment of spins allows for more efficient electron transfer between ions in different oxidation states. This efficient transfer lowers the system's energy and results in a ferromagnetic alignment of the magnetic moments. Double exchange is particularly significant in materials like manganites, where it plays a crucial role in their magnetic properties.

- *Spin-orbit coupling*: Spin-orbit coupling (SOC) is an interaction between an electron's spin and its motion around the nucleus and is shown in **Figure 4(b)**. This interaction is inherently relativistic, arising due to the relativistic corrections to the motion of electrons in an electric field, such as that produced by the nucleus of an atom. SOC plays a crucial role in many physical phenomena and significantly influences the electronic structure, magnetic properties, and behavior of materials. In some materials, the interaction between an electron's spin and its orbital motion around the nucleus contributes to the overall magnetic behavior. This coupling can enhance the magnetic anisotropy of the material, favoring certain directions for the magnetic moments.

- *Mechanism of spin-orbit coupling*: The interaction can be understood by considering an electron moving through the electric field generated by the positively charged nucleus. According to the theory of relativity, an electron moving through an electric field will experience this field as a magnetic field in its rest frame. This perceived magnetic field interacts with the intrinsic magnetic moment of the electron, which is associated with its spin. Mathematically, the spin-orbit coupling interaction is expressed as:

$$H_{SO} = \lambda L \cdot S \tag{8}$$

where H_{SO} is the Hamiltonian representing the spin-orbit interaction, λ is the spin-orbit coupling constant, which depends on the atomic number and the specific electronic environment. L is the orbital angular momentum of the electron. S is the spin angular momentum of the electron. This Hamiltonian describes the energy associated with the interaction between the electron's orbital motion and its spin. The term $L \cdot S$ indicates that the energy depends on the relative orientation of these two vectors.

- *Crystal field effects*: Crystal field theory (CFT) describes how the electric fields of surrounding ligands influence the energy levels of d or f orbitals in a transition metal ion, significantly affecting their magnetic properties. When transition metal ions are placed in a crystal field, the degeneracy of their d or f orbitals is lifted, causing them to split into different energy levels based on the ligand field geometry, such as octahedral, tetrahedral, or square planar arrangements. Crystal field splitting (shown in **Figure 4(b)**) directly impacts the magnetic properties of transition metal complexes by determining the number of unpaired electrons. More unpaired electrons result in stronger paramagnetism, which can be quantified using the magnetic moment (μ). Additionally, the specific splitting patterns in different geometries influence magnetic anisotropy, affecting the material's stability and the directionality of its magnetic properties.

2.4 Magnetic domains and domain wall motion

Magnetic domains are regions within a ferromagnetic material where the magnetic moments are uniformly aligned. The existence of domains minimizes the material's overall energy by reducing the magnetostatic energy associated with stray magnetic fields [14]. Domain wall motion is the process by which the boundaries between domains move in response to an external magnetic field, leading to changes in the material's magnetization. Domains form to reduce the magnetostatic energy of a ferromagnetic material. In the absence of an external field, domains are oriented in such a way that the net magnetic moment of the material is minimized. The boundaries between domains are known as domain walls. These walls are regions where the direction of magnetization changes gradually over a finite distance. There are different types of domain walls, including Bloch walls and Néel walls, depending on the nature of the magnetization transition. When an external magnetic field is applied, domain walls move to align the magnetic moments in the direction of the field. This motion is responsible for the magnetization process in ferromagnetic materials. Domain wall motion can be influenced by various factors, including defects, impurities, and temperature.

3. Microwave-absorbing materials

MAMs are engineered to attenuate microwave radiation by converting electromagnetic energy into heat, which is then dissipated. The schematic of the interaction between EM waves and absorber is shown in **Figure 5**. These materials are critical in a variety of applications, including stealth technology, electromagnetic interference (EMI) shielding, and improving the performance of wireless communication systems. The increasing prevalence of electronic devices and the growing demand for electromagnetic compatibility (EMC) have underscored the importance of developing effective microwave absorbers. Microwave absorbers operate by exploiting dielectric and magnetic losses. Dielectric loss occurs due to the polarization of electric dipoles within the material, while magnetic loss results from the alignment and movement of magnetic domains. The primary goal in designing MAMs is to maximize these losses, thereby enhancing the material's ability to absorb and dissipate microwave energy.

3.1 Types of microwave-absorbing materials

MAMs can be broadly categorized into dielectric absorbers, magnetic absorbers, and composite absorbers. Each type of absorber has distinct characteristics and mechanisms that contribute to microwave absorption.

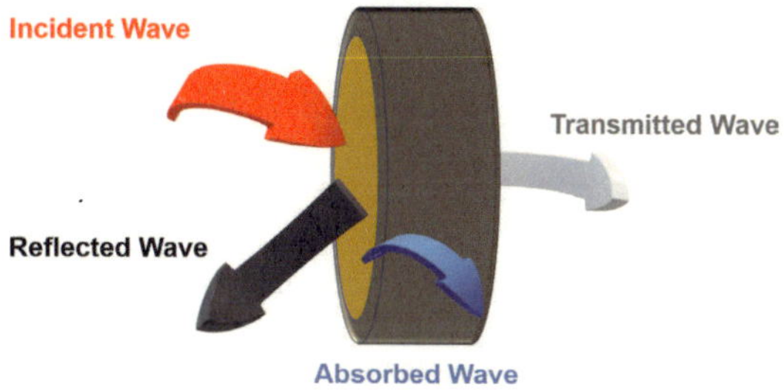

Figure 5.
Schematic diagram of the interaction between EM waves and absorber.

3.1.1 Dielectric absorbers

Dielectric absorbers primarily rely on dielectric losses to attenuate microwave radiation. These materials are typically non-conductive and exhibit high dielectric constants [15]. Key mechanisms include electronic polarization, where the electron cloud is displaced relative to the nucleus under an external electric field, generating an oscillating dipole moment; ionic polarization, involving the displacement of positive and negative ions in an ionic solid, leading to energy loss; and dipole relaxation, where polar molecules align with the field and dissipate energy as they return to random orientation. Common dielectric absorbers, such as barium titanate ($BaTiO_3$) and titanium dioxide (TiO_2), are often combined with conductive fillers or other dielectric materials to enhance absorption properties.

3.1.2 Magnetic absorbers

Magnetic absorbers utilize magnetic losses to absorb microwave energy, leveraging materials that exhibit ferromagnetism, ferrimagnetism, or antiferromagnetism [15]. Key mechanisms include hysteresis loss, arising from the lag between applied magnetic fields and material magnetization; eddy current loss, where induced currents generate opposing magnetic fields and dissipate energy, significant in high-conductivity materials; and natural resonance, involving magnetic dipole resonance with microwave frequencies for maximum absorption. Common magnetic absorbers like iron oxide (Fe_3O_4), cobalt ferrite ($CoFe_2O_4$), and nickel-zinc ferrite ($NiZnFe_2O_4$) can be tailored to specific frequency ranges by adjusting their composition and structure.

3.1.3 Composite absorbers

Composite absorbers combine dielectric and magnetic materials to enhance microwave absorption through synergistic effects [16]. Ferroelectric-ferromagnetic composites balance high dielectric losses from ferroelectric materials with high magnetic losses from ferromagnetic materials, maximizing absorption. Conductive polymer composites, incorporating polymers like polyaniline (PANI) or polypyrrole (PPy) with dielectric or magnetic fillers, offer tunable electrical conductivity and dielectric properties. Carbon-based composites, using fillers such as graphene, carbon nanotubes (CNTs), and carbon black, provide high electrical conductivity and mechanical strength, further enhancing microwave absorption. These designs aim to achieve broadband absorption and superior performance [17].

The development of composite absorbers involves optimizing the composition, morphology, and interfacial interactions of the constituent materials to achieve superior performance across a wide frequency range.

3.2 Challenges in developing effective microwave absorbers

The development of effective microwave-absorbing materials (MAMs) presents several challenges, as discussed below:

i. *Broadband absorption*: Achieving broadband absorption across a wide frequency range, typically from 1 GHz to 18 GHz, is challenging. Materials must exhibit strong dielectric and magnetic losses at multiple frequencies, necessitating precise control over their composition and structure. One approach involves creating composites that combine different types of absorbers to exploit their individual strengths over a wide frequency range. For instance, materials with high dielectric losses at lower frequencies can be integrated with those exhibiting strong magnetic losses at higher frequencies to enhance broadband absorption. Advanced techniques like nano-structuring and doping are often employed to achieve the desired performance [18, 19].

ii. *Thickness and weight*: For applications such as stealth technology and EMI shielding, MAMs must be lightweight and thin, generally ranging from 1 to 3 mm in thickness, with densities around 1–2 g/cm^3. Thinner materials are easier to integrate without adding significant bulk or weight. However, achieving high absorption efficiency in thin materials requires precise control over the material's composition and structure. Techniques like layering different materials, using porous structures, and incorporating lightweight fillers such as carbon nanotubes or graphene can help achieve the desired balance [20, 21].

iii. *Thermal stability*: MAMs are often exposed to high temperatures and harsh conditions, making thermal stability crucial. Materials should maintain their properties up to temperatures of 200°C to ensure reliability. This is particularly important in aerospace and automotive applications where extreme conditions are common. Researchers are developing composites that combine thermally stable polymers with ceramic or metallic fillers to enhance thermal stability while retaining microwave absorption properties [22, 23].

iv. *Mechanical properties*: The mechanical strength and flexibility of MAMs are vital for their integration into various applications. Materials must withstand mechanical stresses without compromising their absorption properties. High tensile strength and flexibility are required, with specific values depending on the application. Advanced composite materials incorporating carbon fibers, aramid fibers, or other high-strength fillers are often used to enhance mechanical properties. Flexible polymers combined with these fillers create materials that are both strong and flexible, suitable for a wide range of applications [24, 25].

v. *Cost and scalability*: The cost-effective production and scalability of MAMs are critical for commercial viability. High-performance materials must be synthesized using economically feasible methods to ensure widespread adoption. Researchers are exploring various synthesis methods to produce high-performance MAMs at lower costs, including using abundant and inexpensive raw materials. Developing bio-waste and e-waste-based microwave absorbers is a promising avenue [26–29]. Scalable manufacturing techniques that can be easily adapted to large-scale production are essential for the commercial success of these materials.

4. Mechanisms of microwave absorption

The effectiveness of MAMs hinges on their ability to convert electromagnetic energy into other forms of energy, typically heat, through various loss mechanisms. These mechanisms can be broadly categorized into dielectric losses, magnetic losses,

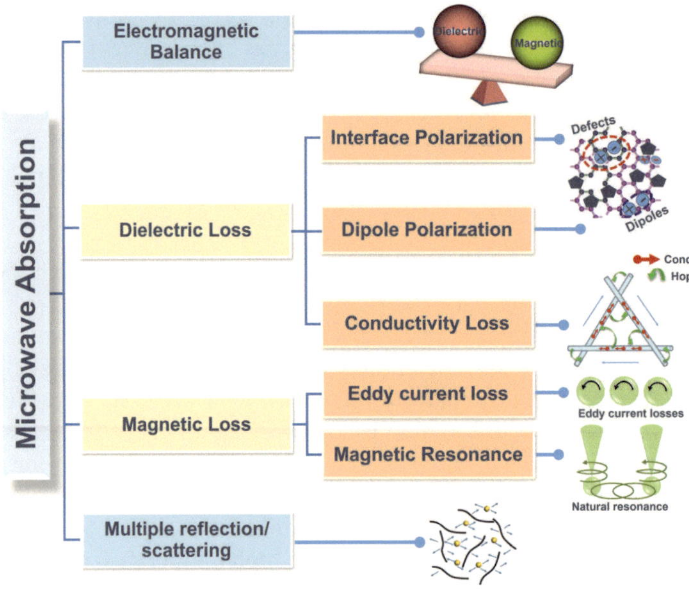

Figure 6.
Schematic of microwave absorption mechanism.

and combined losses in composite materials. Understanding the mechanisms of microwave absorption is critical for designing materials that effectively mitigate electromagnetic radiation. The overall microwave absorption mechanism is shown in **Figure 6.**

4.1 Dielectric loss mechanisms

Electric polarization plays a crucial role in the dielectric loss mechanisms of MAMs. Dielectric loss occurs in materials that are poor conductors of electricity but can support an electrostatic field. It involves the alignment of electric dipoles within the material in response to an external electric field [30]. The primary mechanisms of dielectric loss include:

- *Dipolar relaxation*: In materials with polar molecules, an external electric field aligns the dipoles. When the field changes or is removed, the dipoles reorient themselves back to their original states, dissipating energy as heat. This mechanism is frequency-dependent and is significant in the microwave range [31].

$$\varepsilon'' = \varepsilon_s + \frac{\varepsilon_\infty - \varepsilon_s}{1+(\omega\tau)^2} \tag{9}$$

where ε'' is the imaginary part of the complex permittivity, representing dielectric loss, ε_s is the static permittivity, ε_∞ is the permittivity at infinite frequency, ω is the angular frequency, and τ is the relaxation time.

Different types of polar molecules affect dipolar relaxation in various ways. For example, water molecules exhibit strong dipolar relaxation due to their high dipole moment, making them highly responsive to external electric fields in the microwave frequency range. Polymers with polar side groups, such as polyvinyl alcohol (PVA), also show significant dipolar relaxation. The specific frequency range for these effects depends on the material; water molecules typically respond in the GHz range, while polymer dipoles might respond in the MHz to GHz range depending on their structure and molecular dynamics.

- *Ionic conduction*: In ionic solids, the movement of ions under an electric field causes energy dissipation due to collisions and lattice vibrations. This mechanism is significant at intermediate frequencies and contributes to dielectric loss through the oscillation of ions [32].

$$\sigma = \sigma_0 \exp\left(-\frac{E_a}{kT}\right) \tag{10}$$

where σ is the electrical conductivity, σ_0 is the pre-exponential factor, E_a is the activation energy, k is the Boltzmann constant, and T is the temperature.

Ionic conduction is prominent in materials such as ionic crystals. The mobility of ions and the lattice structure play crucial roles in this mechanism. High ion mobility and a well-ordered lattice structure facilitate ionic conduction, leading to significant dielectric losses. For instance, in ZrO_2 ceramics, the presence of oxygen vacancies enhances ionic mobility, thereby increasing ionic conduction losses.

- *Interfacial polarization (Maxwell-Wagner-Sillars effect)*: In heterogeneous materials, charge accumulation occurs at interfaces between different phases, leading to space-charge polarization. This effect is particularly significant in composite materials and contributes to dielectric loss [33].

$$\varepsilon'' \approx \frac{\sigma}{\varepsilon_0 \omega} \tag{11}$$

where σ is the electrical conductivity, ε_0 is the permittivity of free space, and ω is the angular frequency.

Composite materials enhance interfacial polarization by creating numerous interfaces where charge accumulation can occur. For example, in polymer-ceramic composites, the interfaces between the polymer matrix and ceramic fillers act as sites for space-charge accumulation, significantly contributing to dielectric losses. The heterogeneity of the material, with different phases having varying permittivities and conductivities, is key to maximizing this effect.

4.2 Magnetic loss mechanisms

Magnetic loss occurs in materials with unpaired electrons and is primarily due to the behavior of magnetic domains and their response to an external magnetic field. Magnetic domains are regions within a ferromagnetic material where the magnetic moments are uniformly aligned. The behavior of these domains under an external magnetic field contributes to magnetic loss mechanisms [34]. The key mechanisms include:

- *Hysteresis loss*: Hysteresis loss results from the lag between the applied magnetic field and the magnetization of the material. This lag creates a loop (hysteresis loop) when plotting magnetization versus the magnetic field, indicating energy dissipation. The movement of domain walls in response to an external magnetic field leads to energy dissipation. Domain walls can be pinned by defects, grain boundaries, and impurities, enhancing hysteresis loss [35].

$$P_{\text{hysteresis}} = \mu_0 \int HdM \tag{12}$$

where $P_{\text{hysteresis}}$ is the hysteresis loss, μ_0 is the permeability of free space, H is the magnetic field strength, and M is the magnetization.

Domain wall pinning by defects, grain boundaries, and impurities significantly affects hysteresis loss. In ferromagnetic materials like iron and nickel, impurities and grain boundaries can impede the movement of domain walls, increasing energy dissipation. For instance, the addition of non-magnetic impurities such as sulfur in iron creates pinning sites that enhance hysteresis loss, making these materials more effective for microwave absorption.

- *Eddy current loss:* Induced currents within the material generate opposing magnetic fields, leading to energy dissipation. Eddy current loss is significant in materials with high electrical conductivity. Eddy currents are induced within conductive materials when subjected to a changing magnetic field. These currents generate opposing magnetic fields, leading to energy dissipation [36]:

$$P_{eddy} = \frac{B_{\max}^2 d^2 f^2}{6\rho} \tag{13}$$

where P_{eddy} is the eddy current loss, $B_{\max}$ is the maximum flux density, d is the thickness of the material, f is the frequency of the applied field, and ρ is the electrical resistivity.

Material conductivity and thickness play crucial roles in eddy current losses. High-conductivity materials such as copper generate significant eddy currents, leading to higher losses. To mitigate these losses, materials with lower conductivity or laminated structures are used. For instance, ferrites (low-conductivity ceramics) are employed in microwave absorbers to reduce eddy current losses. Additionally, thinner materials help minimize eddy currents by reducing the path length for current flow.

- *Natural resonance*: Natural resonance occurs when the frequency of the applied magnetic field matches the natural frequency of precession of the magnetic dipoles. The resonance of magnetic dipoles with the applied microwave frequency results in maximum energy absorption. This mechanism is frequency-dependent and contributes significantly to magnetic loss [11]:

$$f_r = \frac{\gamma H}{2\pi} \tag{14}$$

where f_r is the resonance frequency, γ is the gyromagnetic ratio, and H is the magnetic field strength.

The gyromagnetic ratio (γ) and applied magnetic field strength (H) directly influence the resonance frequency (f_r). Higher γ values (as seen in materials like yttrium iron garnet, YIG) and stronger magnetic fields result in higher resonance frequencies, enhancing energy absorption at specific microwave frequencies. Fine-tuning these parameters allows for designing materials with optimal resonance frequencies for targeted applications.

4.3 Combined loss mechanisms in composite materials

Composite materials that integrate dielectric and magnetic components exhibit enhanced microwave absorption through the interplay of various loss mechanisms. This synergy arises from the interaction between different materials and the optimization of their properties. By combining materials with distinct loss mechanisms, researchers can achieve improved broadband absorption [37, 38]. The performance of composite absorbers can be significantly influenced by the relative proportions of different phases, which must be optimized to balance dielectric and magnetic losses for maximum absorption efficiency. Additionally, the interfaces between these phases play a crucial role, enhancing losses through mechanisms such as space-charge polarization and domain wall pinning, which contribute to the overall absorption performance. Incorporating nanomaterials into composites further enhances their properties. Nanomaterials provide a high surface area, tunable electronic properties, and strong interfacial interactions, all of which improve microwave absorption. For example, combining carbon-based materials like graphene or carbon nanotubes with ferrites creates composites that benefit from both the high electrical conductivity and large surface area of carbon-based materials and the magnetic properties of ferrites. This combination leads to enhanced microwave absorption across a wide frequency range due to the synergistic effects of combined dielectric and magnetic losses.

4.4 Losses due to multiple internal reflections and scattering

Losses arising from multiple internal reflections and scattering significantly impact the behavior of electromagnetic waves within various materials. When microwaves encounter interfaces or boundaries between different media, they undergo multiple internal reflections, causing energy to be repeatedly redirected and attenuated within the material. This process leads to losses as the electromagnetic energy is converted into other forms, such as thermal energy. Additionally, scattering occurs when microwaves interact with irregularities or structures within the material, further contributing to energy dissipation. The scattered waves deviate from their original paths, dispersing energy and reducing the overall intensity of the microwave signal. These losses due to multiple internal reflections and scattering are crucial considerations in the design and optimization of microwave-absorbing materials.

4.5 Coupling effects between ferroelectric and ferromagnetic properties

In composite materials with ferroelectric and ferromagnetic phases, magnetoelectric coupling enhances microwave absorption by linking electric and magnetic dipoles. This effect, quantified by the magnetoelectric coupling coefficient $\alpha_{ij} = \frac{\partial P_i}{\partial H_j}$ (α_{ij} is the magnetoelectric coupling coefficient, P_i is the polarization, and H_j is the magnetic field), describes how an electric field affects magnetic properties and vice versa. The interaction between these phases can amplify dielectric and magnetic losses. Specifically, the alignment of electric dipoles can affect magnetic domain movement, leading to increased energy dissipation. Additionally, the interfaces between these phases can create localized states and defects, further boosting both dielectric and magnetic losses.

4.6 Factors influencing microwave absorption

4.6.1 Domain wall motion

Domain walls, the boundaries between regions of uniform polarization or magnetization, move under external fields, causing energy dissipation. In ferroelectrics, domain walls separate regions with different electric polarization orientations, and their motion under an electric field leads to dielectric loss. In ferromagnetics, domain walls separate regions with different magnetization orientations, and their motion under a magnetic field leads to magnetic loss. These interactions with microwave radiation, including resonance, pinning and depinning, and damping mechanisms, significantly contribute to overall microwave absorption in ferroelectric and ferromagnetic materials. Defects such as vacancies, interstitial atoms, and dislocations can pin domain walls, increasing energy dissipation. High defect density in doped ferroelectric materials enhances dielectric loss by increasing pinning sites. In ferromagnetic materials, smaller grain sizes lead to more grain boundaries, which impedes domain wall mobility and increases magnetic loss.

4.6.2 Phase transitions

Phase transitions in ferroic materials, such as ferroelectric and ferromagnetic materials, significantly influence their electric and magnetic properties, thereby affecting microwave absorption. Ferroelectric, ferromagnetic, and multiferroic phase transitions involve changes in crystal structure, magnetic alignment, and simultaneous transitions in electric and magnetic ordering, respectively. For instance, the ferroelectric to paraelectric transition in materials like $BaTiO_3$ causes an abrupt change in the dielectric constant, impacting the material's ability to absorb microwave energy. Similarly, the ferromagnetic to paramagnetic transition in materials like $NiFe_2O_4$ alters magnetic properties, influencing magnetic loss mechanisms. These phase transitions can thus alter dielectric and magnetic properties, affecting microwave absorption capabilities. Additionally, temperature, pressure, and external fields can influence phase transitions, providing a means to tune material properties for optimal microwave absorption. Kumar *et al.* [39] studied the effect of heat treatment on morphology and microwave absorption of SiC particles. Through ball milling and heat treatment, the researchers transformed these particles into spherically shaped nano-size SiC and a one-dimensional SiC/Si_3N_4 micro-whiskers composite system. The results indicated a significant enhancement in dielectric properties for both the spherically shaped nano-size SiC and 1D SiC/Si_3N_4 micro-whiskers when compared to the original particles. Microwave absorption performance showed notable improvement in 8–12 GHz frequency range for the shape-modified SiC particles. A similar study was carried out by Wei *et al.* [40] where they explored the impact of heat treatment on absorption properties of flaky CI powder produced through high-energy ball milling. The study aimed to optimize microstructure and enhance absorption by analyzing the influence of pre-heating at various times and temperatures. Microstructural analysis revealed finer grain size and increased flattening ratio in ball-milled material. Pre-heating increased the length-to-diameter ratio, improving magnetic permeability and absorption performance. Optimal pre-heating parameters were identified as 200°C for 2 hours, achieving a real part of permeability up to 3.20 at 2 GHz and an imaginary part of 1.61 at 6.2 GHz.

4.6.3 Structure: Property relationship

The arrangement of atoms within a crystal lattice affects the electronic and magnetic properties, influencing microwave absorption. The shape and size of particles or grains impact surface area and interaction with electromagnetic waves, with nanostructured materials often exhibiting enhanced absorption [33]. Defects and impurities introduce localized states that contribute to dielectric loss and alter magnetic properties, affecting microwave absorption [30]. Oyharçabal *et al.* [41] explored the impact of polyaniline (PANI) morphology on absorption properties within epoxy polyaniline composites. The study synthesized various PANI structures—globular, fibrillar, and flake-like—incorporated into epoxy resin. Increasing the aspect ratio of PANI resulted in a lowered electrical percolation threshold (from 3.9 to 1.3 vol.%) and increased electrical conductivity. Specifically, the inclusion of high aspect ratio flake-like PANI significantly enhanced microwave absorption properties (2.4–8.8 GHz). Composites with flake-like PANI exhibited a doubled ε'' compared to globular and fibrillar PANI. Yang *et al.* [42] studied the influence of Fe particle size and shape on composite absorbers' absorbing properties. Through controlled mechanical milling of Fe powders in two batches, they produced polymer composites featuring flake-shaped Fe particles. These composites, with 40 wt.% of such particles, displayed increased magnetic permeability and dielectric permittivity. The research unveiled a substantial increase in both ε' and μ' as the milling time and particle size grew, attributed to enhanced space-charge polarization. Guo *et al.* [43] studied the effect of shape anisotropy, where the spherical CI particles were transformed into planar, anisotropic particles through milling. During fabrication, an external magnetic field strategically aligned these PACIs within a paraffin composite, resulting in the orientation of shape anisotropy field. This alignment improved complex permeability while simultaneously reducing complex permittivity, leading to good impedance matching. The produced absorber displayed an absorption bandwidth of 1.4 GHz for a thickness of 3.25 mm. These studies highlight the potential of tailoring particle shape and orientation for designing microwave absorbers.

4.6.4 Influence of microstructure

The influence of microstructure on microwave absorption is significant. Smaller grain sizes enhance dielectric loss through space-charge polarization and reduce eddy current losses by increasing electrical resistivity. Grain boundaries act as barriers to domain wall motion, contributing to magnetic hysteresis loss and affecting dielectric properties. Porosity scatters electromagnetic waves, enhancing absorption but potentially weakening mechanical strength. Huang *et al.* [44] investigated the impact of pore morphology of porous carbons (PCs) on microwave absorption. Using a sol–gel process and freeze-drying, they created PCs with adjustable morphologies and pore sizes. Results showed that, under similar graphitization conditions, pore morphology had a dominant effect on dielectric loss. The sample with cage-like pores exhibited maximum values in real and imaginary parts of complex permittivity within the 8.2–12.4 GHz range, indicating enhanced dielectric loss capacity. The study proposed a mechanism to explain how pore morphologies influence microwave absorption performance. Liu *et al.* [45] studied the influence of the size of Fe_3O_4 nanospheres on absorption properties. They utilized the solvothermal method to create Fe_3O_4 nanoparticles with adjustable sizes (20–250 nm). The results demonstrated a significant correlation between granular size, permeability, and absorption properties. Notably, Fe_3O_4 nanoparticles with a 20 nm size, aided by magnetic stirring, exhibited enhanced absorption compared to larger counterparts produced without magnetic stirring. They concluded that Fe_3O_4 nanoparticles with a size under 100 nm are potential candidates for microwave absorption.

4.6.5 Optimization through compositional control

Optimization through compositional control is crucial for enhancing microwave absorption properties. Introducing dopants can modify electronic and magnetic properties, increasing the dielectric constant or enhancing permeability. Adjusting the relative proportions of different phases can balance dielectric and magnetic losses, optimizing overall properties. Combining different materials creates hybrid materials with superior absorption properties through synergy between components. Incorporating nanomaterials into nanocomposites significantly improves properties by offering high surface area, tunable electronic properties, and strong interfacial interactions. Zhao *et al.* [46] examined the impact of varying Ni contents on the absorbing properties of FeNi nanopowders. The magnetic loss of $Fe_{100-x}Ni_x$ nanopowders dominated within the 2–14 GHz frequency range. As the Ni content increased, the peak magnetic loss shifted to higher frequencies. Singh *et al.* [47] investigated the effect of Zn metal particle dispersion in SiC on absorption characteristics. Their findings revealed an enhancement in EM wave absorption within the 8–18 GHz frequency range due to the dispersion of Zn particles in the SiC matrix.

Jamwal *et al.* have made significant strides in the field of MAMs by exploring various factors affecting their performance. Their research reveals that adjusting particle sizes and incorporating multi-walled carbon nanotubes (MWCNTs) into cobalt (Co) absorbers enhances microwave absorption through improved interfacial polarization and multiple reflections, achieving an effective bandwidth of up to 10 GHz [17]. In their study of carbonyl iron/barium hexaferrite (CI/BFO) nanocomposites prepared via high-energy ball milling, they found that the modified morphology and complex electromagnetic interactions within the composites lead to an impressive absorption range with an effective bandwidth of 15.12 GHz when optimized with a triple-layer design [16]. Their investigation into gamma radiation's effect on lithium-substituted nickel ferrite (LNFO) shows that gamma exposure induces structural changes and increases free electron density, which enhances dielectric and magnetic losses, resulting in a bandwidth of 3.8 GHz [48]. Additionally, Jamwal *et al.* explored the impact of ball milling with heterogeneous ball sizes, demonstrating that this method produces finer particles and improves absorption characteristics in cobalt (Co) through increased surface area and enhanced electromagnetic interaction, achieving a bandwidth of 9.65 GHz [49]. Collectively, these studies offer valuable insights into the mechanisms of microwave absorption and innovative approaches for optimizing materials for high-performance applications.

By understanding and optimizing these mechanisms, researchers can design advanced MAMs that effectively mitigate electromagnetic radiation, meeting the demands of modern technology. The following sections will explore practical applications and future research directions in this field.

5. Applications and practical implications

MAMs have a wide range of applications due to their ability to mitigate electromagnetic interference (EMI) and enhance the performance of electronic devices. EMI refers to unwanted disturbances caused by external electromagnetic fields that can degrade signal quality, cause data loss, and lead to malfunctions in electronic devices. Such interference can stem from natural phenomena, human-made equipment, or the devices themselves, particularly in environments with multiple electronic devices operating in close proximity. As electronic circuits become more complex and miniaturized, and wireless communication technologies proliferate, susceptibility to EMI has grown. To ensure reliable operation and electromagnetic compatibility (EMC), effective EMI shielding is crucial. MAMs play a key role in this by absorbing incident electromagnetic waves and converting them into heat, thereby reducing wave intensity, reflection, and interference. They are designed to operate across a broad frequency range to address diverse EMI sources, with composite materials that combine dielectric and magnetic losses providing enhanced broadband absorption. MAMs can be applied as coatings, films, or structural components in enclosures, cables, and connectors to offer comprehensive EMI protection. Beyond shielding, MAMs significantly improve the efficiency and performance of various electronic devices. For sensors, MAMs enhance performance by reducing EMI and improving signal integrity. In radar and imaging systems, they minimize unwanted reflections and increase detection accuracy, while in automotive radar, they bolster the reliability of collision avoidance and adaptive cruise control systems. In antennas, MAMs improve performance by reducing backscattering and surface currents, leading to better radiation patterns and increased gain. They also reduce mutual coupling between elements in phased array antennas, enhancing beamforming capabilities and signal-to-noise ratio, and are used in anechoic chambers for interference-free antenna testing. In communication devices, MAMs play a crucial role in reducing interference and multipath effects, thereby enhancing signal quality. In mobile phones, they lower specific absorption rate (SAR) and minimize signal loss, which improves call quality and battery life. For Wi-Fi routers, MAMs reduce interference from nearby devices, resulting in more stable and faster internet connections. Overall, MAMs are integral to optimizing electronic device performance and ensuring effective EMI management.

5.1 Prospects for further research and development

The ongoing research and development in the field of MAMs is focused on addressing current challenges and exploring new possibilities. Some of the promising directions include:

i. Developing materials with enhanced broadband absorption capabilities remains a key goal. Researchers are exploring new compositions, structures, and fabrication techniques to achieve high absorption efficiency across a wide frequency range.

ii. Ensuring the scalability and cost-effectiveness of advanced MAMs is crucial for their widespread adoption. Research efforts are aimed at developing economically feasible synthesis methods and reducing material costs.

iii. The integration of MAMs with emerging technologies, such as 5G, the Internet of Things (IoT), and autonomous vehicles, presents new opportunities and challenges. Tailoring absorbers to meet the specific requirements of these technologies is an active area of research.

iv. Addressing environmental and health concerns associated with MAMs is important for their sustainable development. Researchers are investigating eco-friendly materials and assessing the potential health impacts of long-term exposure to microwave absorbers.

6. Conclusion

This chapter explores the mechanisms behind microwave absorption, focusing on the interplay between electric polarization and ferromagnetic properties. Electric polarization involves aligning electric dipoles under an external field, while ferromagnetism results from the parallel alignment of magnetic moments. Key dielectric losses include electronic, ionic, dipolar, and space-charge polarization. Magnetic losses arise from domain wall motion, hysteresis, eddy currents, and natural resonance. Composite absorbers combine dielectric and magnetic materials, leveraging both to enhance microwave absorption. This synergy allows for broadband absorption across a wide frequency range, which is crucial for applications in stealth technology, electromagnetic interference shielding, and wireless communication systems. Understanding these mechanisms and the synergy between dielectric and magnetic properties guides the development of next-generation MAMs. Hybrid materials with nanostructures like graphene or carbon nanotubes promise superior performance. Emerging technologies such as 5G and advanced stealth systems will benefit from these advancements. Future research should explore new material systems with unique dielectric and magnetic properties, novel ferroelectric and ferromagnetic materials, and hybrids with nanomaterials. Studies on microstructure and interfaces in microwave absorption are also needed. Advanced characterization techniques, like in situ methods, can reveal microstructural features influencing losses. Theoretical modeling and simulation can enhance understanding. Investigating phase transitions and environmental effects on material properties is crucial. Developing scalable, cost-effective synthesis methods is essential for commercial viability. Additionally, multifunctional materials combining microwave absorption with mechanical strength and thermal conductivity have broad applications in aerospace and consumer electronics.

Acknowledgements

Udeshwari Jamwal acknowledges Indian Institute of Technology Roorkee and Shivam Kumar Mittal and Deepanshu Kaneria acknowledge Council of Scientific and Industrial Research (CSIR) for research funding. The authors acknowledge the use of ChatGPT for language editing in the chapter.

Conflict of interest

The authors declare no conflict of interest.

Author details

Udeshwari Jamwal*, Shivam Kumar Mittal and Deepanshu Keneria
Department of Physics, Indian Institute of Technology Roorkee, Roorkee, India

*Address all correspondence to: udeshwarimpy020@gmail.com

References

[1] Ahlbom A, Feychting M. Electromagnetic radiation: Environmental pollution and health. British Medical Bulletin. December 2003;**68**(1):157-165. DOI: 10.1093/BMB/LDG030

[2] Ahmad H et al. Stealth technology: Methods and composite materials—A review. Polymer Composites. 2019;**40**(12):4457-4472. DOI: 10.1002/pc.25311

[3] Khan J, Duan W, Sherbaz S. Radar cross section prediction and reduction for naval ships. Journal of Marine Science and Application. 2012;**11**(2):191-199. DOI: 10.1007/S11804-012-1122-5/METRICS

[4] Green M, Chen X. Recent progress of nanomaterials for microwave absorption. Journal of Materials. 2019;**5**(4):503-541. DOI: 10.1016/j.jmat.2019.07.003

[5] Yusuf JY et al. Recent advances and prospect of cobalt based microwave absorbing materials. Ceramics International. December 2020;**46**(17):26466-26485. DOI: 10.1016/J.CERAMINT.2020.07.244

[6] Saeed M, Haq RSU, Ahmed S, Siddiqui F, Yi J. Recent advances in carbon nanotubes, graphene and carbon fibers-based microwave absorbers. Journal of Alloys and Compounds. January 2024;**970**:172625. DOI: 10.1016/J.JALLCOM.2023.172625

[7] Zhao Z et al. Advancements in microwave absorption motivated by interdisciplinary research. Advanced Materials. 2023;**36**(4):2304182. DOI: 10.1002/adma.202304182

[8] Gao T et al. Sub-nanometer Fe clusters confined in carbon nanocages for boosting dielectric polarization and broadband electromagnetic wave absorption. Advanced Functional Materials. 2022;**32**(31):1-10. DOI: 10.1002/adfm.202204370

[9] Yeswanth IVS, Jha K, Bhowmik S, Kumar R, Sharma S, Ilyas RA. Recent developments in RAM based MWCNT composite materials: A short review. Functional Composites and Structures. 2022;**4**(2):024001. DOI: 10.1088/2631-6331/ac5730

[10] Kao KC. Electric polarization and relaxation. In: *Dielectric Phenomena in Solids*. San Diego, California, USA: Elsevier Academic Press; 2004. pp. 41-114

[11] Kittel C. On the theory of ferromagnetic resonance absorption. Physics Review. 1948;**73**(2):155. DOI: 10.1103/PhysRev.73.155

[12] Mittal SK et al. Enhanced sensitivity in capacitive temperature sensors through synergistic relaxor/antiferroelectric composites. IEEE Transactions on Dielectrics and Electrical Insulation. 2024;**31**(3):1119-1126. DOI: 10.1109/TDEI.2023.3344690

[13] Goodenough JB. Summary of losses in magnetic materials. IEEE Transactions on Magnetics. September 2002;**38**(5) Suppl. II:3398-3408. DOI: 10.1109/TMAG.2002.802741

[14] Hajalilou A, Tavakoli M, Parvini E, Introduction to Magnetic Materials, in Magnetic Materials. First published: 22 September. Weinheim, Germany: Wiley-VCH; 2022. DOI: 10.1002/9783527840762.ch1

[15] Pozar DM. *Microwave Engineering*. John Wiley & Sons, Inc.; 2012

[16] Jamwal U, Singh D, Yadav KL. Development of CI/BFO nanocomposite for efficient microwave absorption in low frequency range of L, S, C, and X – Bands. Journal of Alloys and Compounds. Hoboken, New Jersey, USA: John Wiley & Sons, Inc; 2023;**968**:171798. DOI: 10.1016/j.jallcom.2023.171798

[17] Jamwal U, Singh D, Yadav KL. Effect of particle size and MWCNTs content on microwave absorption characteristics of cobalt. IEEE Transactions on Magnetics. 2022;**58**(10):2200516. DOI: 10.1109/TMAG.2022.3199582

[18] Padhy S, De A, Debata RR, Meena RS. Design, characterization, and optimization of a multilayer U-type hexaferrite-based broadband microwave absorber. IEEE Transactions on Electromagnetic Compatibility. 2018;**60**(6):1734-1742. DOI: 10.1109/TEMC.2018.2805364

[19] Yang X et al. Constructing three-dimensional reticulated carbonyl iron/carbon foam composites to achieve temperature-stable broadband microwave absorption performance. Carbon N. Y. 2022;**188**:376-384. DOI: 10.1016/J.CARBON.2021.12.044

[20] Qu Z et al. An ultra-thin ultra-broadband microwave absorber for radar stealth. Advanced Composites and Hybrid Materials. 2022;**5**(3):1778-1785. DOI: 10.1007/S42114-022-00429-Y

[21] Zhou C, Wu C, Liu D, Yan M. Metal-organic framework derived hierarchical Co/C@V2O3 hollow spheres as a thin, lightweight, and high-efficiency electromagnetic wave absorber. *Chemistry–A European Journal*. 2019;**25**(9):2234-2241. DOI: 10.1002/CHEM.201805565

[22] Ghosh SK et al. Carbon nanotubes and carbon nanofibers based co-continuous thermoplastic elastomeric blend composites for efficient microwave shielding and thermal management. Composites. Part A, Applied Science and Manufacturing. 2022;**161**:107118. DOI: 10.1016/J.COMPOSITESA.2022.107118

[23] Gandhi N, Singh K, Ohlan A, Singh DP, Dhawan SK. Thermal, dielectric and microwave absorption properties of polyaniline–CoFe2O4 nanocomposites. Composites Science and Technology. 2011;**71**(15):1754-1760. DOI: 10.1016/J.COMPSCITECH.2011.08.010

[24] Al-Mattarneh H, Dahim M. Physical and mechanical properties of microwave absorber material containing micro and nano barium ferrite. Advanced Materials Letters. 2019;**10**(4):259-262. DOI: 10.5185/AMLETT.2019.2226

[25] Kala T, Maharshi K, Patel S, Panwar R. Electromagnetic and mechanical characterization of iron reinforced natural fiber composites for microwave absorbing applications. Advanced Composite Materials. 2021;**30**(6):559-569. DOI: 10.1080/09243046.2021.1904345

[26] Mishra SP, Nath G, Mishra P. Ultrasonically synthesized dielectric microwave absorbing material from coconut coir dust. Waste and Biomass Valorization. 2020;**11**(4):1481-1490. DOI: 10.1007/S12649-018-0478-4

[27] Qi G et al. Lightweight Fe3C@Fe/C nanocomposites derived from wasted cornstalks with high-efficiency microwave absorption and ultrathin thickness. Advanced Composites and Hybrid Materials. 2021;**4**(4):1226-1238. DOI: 10.1007/S42114-021-00368-0

[28] Zhao H et al. Biomass-derived porous carbon-based nanostructures

for microwave absorption. Nano-Micro Letters. 2019;**11**(1):1-17. DOI: 10.1007/S40820-019-0255-3

[29] Panwar R, Agarwala V, Singh D. A cost effective solution for development of broadband radar absorbing material using electronic waste. Ceramics International. 2015;**41**(2):2923-2930. DOI: 10.1016/J.CERAMINT.2014.10.118

[30] Zhang M et al. A multifunctional stealthy material for wireless sensing and active camouflage driven by configurable polarization. Journal of Materials Science and Technology. 2023;**132**:42-49. DOI: 10.1016/J.JMST.2022.05.046

[31] Wang S, Li Q, Hu K, Liu Q, Liu X, Kong X. Activating microwave absorption performance by reduced graphene oxide-borophene heterostructure. Composites. Part A, Applied Science and Manufacturing. 2020;**138**:106033. DOI: 10.1016/J.COMPOSITESA.2020.106033

[32] Kuang J, Jiang P, Ran F, Cao W. Conductivity-dependent dielectric properties and microwave absorption of Al-doped SiC whiskers. Journal of Alloys and Compounds. 2016;**687**:227-231. DOI: 10.1016/J.JALLCOM.2016.06.168

[33] Qin M, Zhang L, Wu H. Dielectric loss mechanism in electromagnetic wave absorbing materials. Advancement of Science. 2022;**9**(10):2105553. DOI: 10.1002/ADVS.202105553

[34] Pardavi-Horvath M. Microwave applications of soft ferrites. Journal of Magnetism and Magnetic Materials. 2000;**215**:171-183. DOI: 10.1016/S0304-8853(00)00106-2

[35] Ruiz-Perez F, López-Estrada SM, Tolentino-Hernández RV, Caballero-Briones F. Carbon-based radar absorbing materials: A critical review. Journal of Science: Advanced Materials and Devices. 2022;**7**(3):100454. DOI: 10.1016/j.jsamd.2022.100454

[36] Wang YY, Song-Yang WJ, Sun K, Dai DX, Yan Z, Li M. Highly enhanced microwave absorption for carbon nanotube/barium ferrite composite with ultra-low carbon nanotube loading. Journal of Materials Science and Technology. 2022;**102**:115-122. DOI: 10.1016/J.JMST.2021.06.032

[37] Naito Y, Suetake K. Application of ferrite to electromagnetic wave absorber and its characteristics. *IEEE Transactions on Microwave Theory and Techniques*. 1971;**19**(1):65-72. DOI: 10.1109/TMTT.1971.1127446

[38] Griffiths DJ, Inglefield C. Introduction to electrodynamics. American Journal of Physics. 2005;**73**(6):574-574. DOI: 10.1119/1.4766311

[39] Kumar A, Singh S, Singh D. Effect of heat treatment on morphology and microwave absorption behavior of milled SiC. Journal of Alloys and Compounds. 2019;**772**:1017-1023. DOI: 10.1016/j.jallcom.2018.09.136

[40] Wei H et al. Influence of heat treatment on the microwave absorption properties of flaky carbonyl iron powder. *International Journal of Lightweight Materials and Manufacture*. 2020;**3**(3):258-264. DOI: 10.1016/j.ijlmm.2020.02.001

[41] Oyharçabal M, Olinga T, Foulc MP, Lacomme S, Gontier E, Vigneras V. Influence of the morphology of polyaniline on the microwave absorption properties of epoxy polyaniline composites. Composites Science and Technology. 2013;**74**:107-112. DOI: 10.1016/j.compscitech.2012.10.016

[42] Bin Yang R, Liang WF, Choi ST, Lin CK. The effects of size and shape

of iron particles on the microwave absorbing properties of composite absorbers. IEEE Transactions on Magnetics. 2013;**49**(7):4180-4183. DOI: 10.1109/TMAG.2013.2239973

[43] Guo C, Yang Z, Shen S, Liang J, Xu G. High microwave attenuation performance of planar carbonyl iron particles with orientation of shape anisotropy field. Journal of Magnetism and Magnetic Materials. 2018;**454**:32-38. DOI: 10.1016/J.JMMM.2018.01.025

[44] Huang Y, Wang Y, Li Z, Yang Z, Shen C, He C. Effect of pore morphology on the dielectric properties of porous carbons for microwave absorption applications. Journal of Physical Chemistry C. 2014;**118**(45):26027-26032. DOI: 10.1021/jp506999k

[45] Liu J, Zeng M, Yu R, Liu X, Zhu M. Size influence to the high-frequency properties of granular magnetite nanoparticles. IEEE Transactions on Magnetics. 2014;**50**(11):1-4. DOI: 10.1109/TMAG.2014.2331061

[46] Zhao H, Zhu Z, Xiong C, Zheng X, Lin Q. The influence of different Ni contents on the radar absorbing properties of FeNi nano powders. RSC Advances. 2016;**6**(20):16413-16418. DOI: 10.1039/c5ra27179g

[47] Singh S, Shukla S, Kumar A, Singh D. Influence of Zn dispersion in SiC on electromagnetic wave absorption characteristics. Journal of Alloys and Compounds. 2018;**738**:448-460. DOI: 10.1016/j.jallcom.2017.12.190

[48] Jamwal U, Singh D, Yadav KL. An experiment to observe the effect of gamma radiation on microwave absorption properties of Li-substituted nickel ferrite. Journal of Materials Science: Materials in Electronics. 2023;**34**(22):1643. DOI: 10.1007/s10854-023-11060-y

[49] Jamwal U, Singh D, Yadav KL. Unveiling ball size heterogeneity effect on microwave absorption properties of cobalt. Ceramics International. Part A, 1 September 2024;**50**(17):30039-30050. DOI: 10.1016/j.ceramint.2024.05.301

Chapter 5

Anhysteretic Magnetization in Ferromagnetics

Sergey G. Sandomirski

Abstract

Anhysteretic magnetization involves applying a strong alternating magnetic field to a material simultaneously with a constant magnetizing field. The alternating field is then gradually reduced to zero. The initial magnetization curve is replaced by an anhysteretic magnetization curve. Its use sharply increases the sensitivity of magnetic elements. But there is no analytical description of such magnetization. In the chapter, to simplify the determination of the anhysteretic magnetization curve of a ferromagnetic material, a formula is developed using the results of measuring its saturation magnetization M_s, remanent magnetization M_r, and coercive force H_c, which can be measured according to standard techniques with a minimum error. The developed formula is used to determine the internal N_i coefficient of material demagnetization arising due to inhomogeneity's in its structure. The effect of magnetic properties of different materials on N_i has been analyzed. The developed formula is also used to determine the anhysteretic magnetization curve of a ferromagnetic body and to analyze the relationship between the initial and maximum magnetic susceptibilities of a material under normal and anhysteretic magnetization. The results obtained are important for electrical engineering and magnetic structural analysis.

Keywords: ferromagnetic steels, magnetic properties, anhysteretic magnetization, internal magnetization factor, magnetic hysteresis loop, saturation magnetization, remanent magnetization, coercive force

1. Introduction

Structural and electrical steels are typically ferromagnetic materials (**Figure 1**). Their magnetization is characterized (**Figure 1**) by the normal magnetization curve $M(H)$ (geometric location of the vertices of symmetrical magnetic hysteresis loops, which are obtained at successively increasing maximum values of the magnetic field strength [1]) and the Stoletov curve (graph $\chi(H)$ of the dependence of the magnetic susceptibility χ on the magnetic field strength H in a ferromagnetic [2]).

The magnetic parameters whose measurement results are given in Ref. and scientific literature for almost all steels [3–5] are: coercive force H_c, saturation magnetization M_s, and remanent magnetization M_r. If the requirements of [2] are met, the

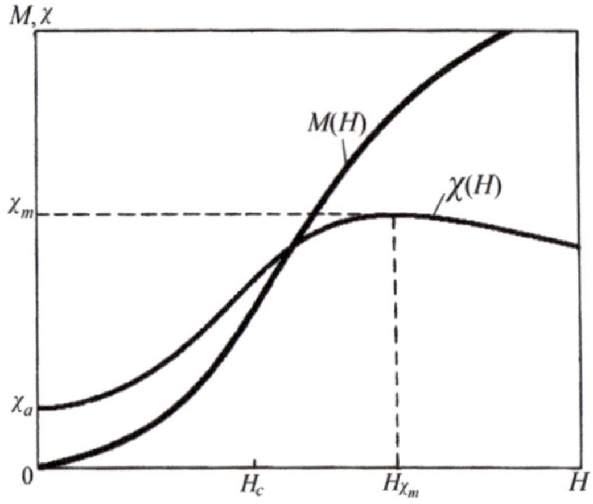

Figure 1.
Normal magnetization curve M(H) and Stoletov curve χ(H) of steel.

relative error δ of measuring H_c of ferromagnetic materials does not exceed $\pm 2\%$, M_s and M_r – $\pm 3\%$, and δ of measuring the ratio $K_r = M_r/M_s$ ("rectangularity factor") when measuring M_s and M_r on the same apparatus does not exceed $\pm 1\%$ (errors in determining the cross-sectional area of the sample and the galvanometer constant are excluded in this case) [6]. K_r and the ratio $\xi = H_c/M_s$ are recommended for use in magnetic structural analysis [7]. Formulas for describing the normal magnetization curve and the Stoletov curve of steels under normal magnetization based on the results of H_c, M_s, and M_r measurements were developed and substantiated in Refs. [8, 9].

Anhysteretic magnetization of the material eliminates the influence of the hysteresis phenomenon on the magnetization of the ferromagnetic material [1, 10–13]. In this process, simultaneously with the main permanent magnetizing field of intensity H, the material is subjected to an additional alternating magnetic field, which brings the material to technical saturation. At each value of H of the constant magnetic field, the amplitude of the alternating magnetic field is smoothly decreased from maximum to zero. The measured values of the magnetization M of the material after each such operation are presented as usual as a function of H. In this case, instead of the curve of initial magnetization [1, 12], the anhysteretic magnetization curve is obtained—geometric location of the magnetization curve points obtained by superimposing an alternating magnetic field with decreasing to zero amplitude on a monotonically increasing permanent magnetic field H [1, 12]. (The term "ideal magnetization curve" used in a number of papers according to reference [1] is "inadmissible for use"). The curve of anhysteretic magnetization refers to the curves of ground states. These are the states that, under given external conditions, have the lowest free energy, that is, are thermodynamically most stable [11, 12]. An additional alternating field leads to the establishment of an equilibrium state. The constant field is not able to make the transition to this equilibrium state independently due to the presence of potential barriers (in intermediate transition states) [11]. Those irreversible magnetization processes that cannot be completed under the action of a weak constant field are also involved in creating the magnetization of the new equilibrium state.

The peculiarity of the anhysteretic magnetization curve is the absence of an inflection point and a small angle α of the curve inclination to the ordinate axis at the point $H = 0$ (**Figure 2**).

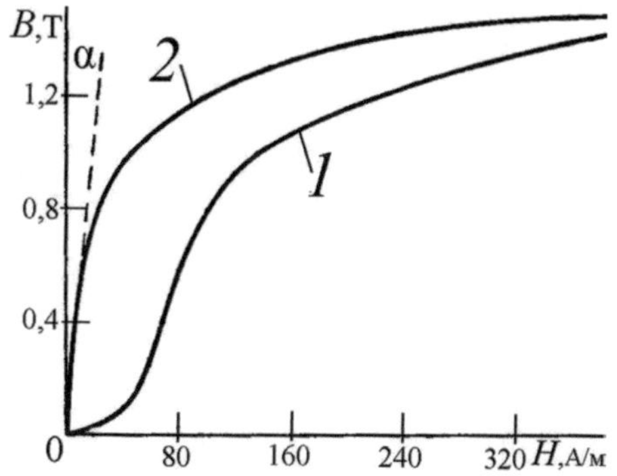

Figure 2.
Basic (1) and anhysteretic (2) magnetization curves of technically pure iron according to reference [12].

The use of anhysteretic magnetization dramatically increases the sensitivity of measuring systems containing magnetic elements [11]. Therefore, an analytical description of the magnetization change on the anhysteretic magnetization curve and the magnetic susceptibility on the Stoletov curve of ferromagnetic materials during their anhysteretic magnetization is important for solving electrical engineering problems.

According to the tangent of the angle α of inclination of the anhysteretic magnetization curve to the ordinate axis at the point $H = 0$ (**Figure 1**), the internal coefficient N_i of demagnetization of the material is determined, which arises due to stresses and heterogeneities in its structure. Therefore, N_i allows us to reliably judge about changes in the structure and mechanical properties of the material [13]. This explains the importance of its determination for nondestructive magnetic analysis of the structure of ferromagnetic materials.

The aim of the chapter is to develop an analytical expression to describe the change in the magnetization M and magnetic susceptibility χ of ferromagnetic materials during their anhysteretic magnetization in terms of the intensity H of the magnetizing field, the coercive force H_c, the saturation magnetization M_s, and the remanent magnetization M_r measured at the saturation hysteresis loop of the material. Development on this basis of a formula for calculating the internal coefficient N_i of demagnetization of ferromagnetic materials. Analysis of the effect of heat treatment of steel on its N_i and Stoletov curve in anhysteretic magnetization. Comparison of maximum magnetic susceptibilities of steel under commutation and anhysteretic magnetization. Development of the analytical dependence for the anhysteretic magnetization curve of a ferromagnetic body.

2. Analyzing the features of measuring the anhysteretic magnetization curve

Measurement [14–16] of the anhysteretic magnetization curve of a ferromagnetic material involves the necessity to carry out a full cycle of specific magnetic actions on the material with numerous (and inaccurate) measurements of its magnetic state during these actions [11, 13, 16]. Another problem is that the experimental curve may not be smooth due to noise introduced during measurements and digitization of their data [13]. Therefore, even its physical description is difficult [17, 18].

The direct application of automated devices [19, 20] for measuring anhysteretic magnetization curve for recording hysteresis loops and their parameters is impossible. This is hindered by the specific magnetic influence on the material. The requirement [1] to measure the anhysteretic magnetization curve on a thermally demagnetized material is due to the need to ensure the repeatability of measurements in weak magnetic fields. But heating a material to Curie temperature can change its magnetic properties obtained from previous thermal treatment. Therefore, experimental studies of the effect of some thermal treatments of the material on the anhysteretic magnetization curve are impossible.

Special devices for measuring the anhysteretic magnetization curve [14, 21] are complicated and do not have high performance due to the need to observe the permissible rate of change of the alternating magnetic field acting on the material [11]. Metrological support of anhysteretic magnetization curve measurements by known automated devices [14–16, 19–21] is not reported. Therefore, to analyze the anhysteretic magnetization curve of a ferromagnetic material, its modeling is used.

3. Analysis of methods for modeling the anhysteretic magnetization curve of ferromagnetic material

Several empirical formulae have been proposed to model the anhysteretic magnetization curve: rational, logarithmic, exponential, inverse arctangent functions [11, 22], and Jiles-Atherton models [23–25], implemented in finite element modeling software. But they do not use the basic magnetic parameters of the material, but empirical results of measuring points (M, H) on the anhysteretic magnetization curve. Analytical functions for the exact description of the whole nonlinear process of anhysteretic magnetization of ferromagnetic materials have not been obtained. Spline interpolation methods are commonly used to smooth anhysteretic magnetization curves. They are combined with extrapolation methods to solve the saturation approximation problem, facing the problem of obtaining physically reasonable results [14, 23, 26].

It was shown in Refs. [10, 27–29] that the anhysteretic magnetization curve of a ferromagnetic material can be calculated as the average curve of the descending and ascending branches of the measured saturation loop of its magnetic hysteresis. In Ref. [27], a method for determining the anhysteretic magnetization curve of a material is proposed, by which the saturation hysteresis loop of the material is measured on a sample of the material under investigation. Then, a straight line parallel to the abscissa axis is drawn through the points A and D (**Figure 3**) with the same induction value on the descending and ascending branches of the measured magnetic hysteresis loop. Determine the magnetic field values at these points. Determine half of the sum of these magnetic field values (taking into account their sign). Its value serves as the value of the magnetic field of point C on the desired anhysteretic magnetization curve of the material. The coordinates of other points of the anhysteretic magnetization curve of the material are set in the same way (**Figure 3**). The method [27] is based on the fact that the anhysteretic magnetization curve is close to the line passing through the midpoints of the horizontal chords of the hysteresis loop corresponding to saturation [10].

In Ref. [27], this position was verified on toroidal samples of CT 80, 60SiCR7, and 34CrNiMo6 steels. The verification showed that the discrepancy between the values of the anhysteretic magnetization curve determined by the specified method and the

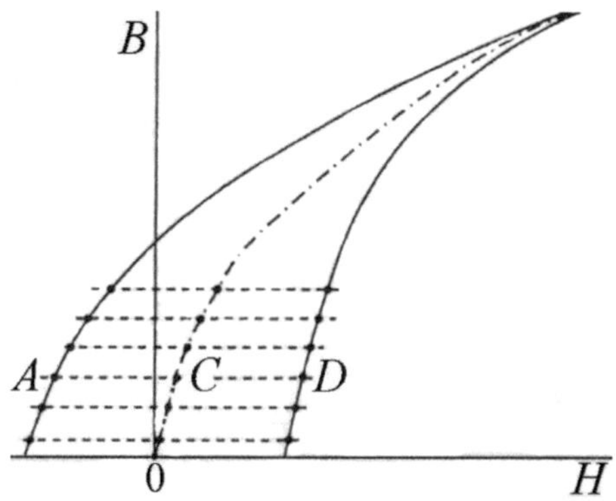

Figure 3.
Schematic of using the results of measuring points on the dependence of induction B on the effective magnetic field H on the descending (A) and ascending (D) branches of the hysteresis loop used to determine points (C) of the anhysteretic magnetization curve according to reference [27].

classical method (under the simultaneous action of a constant magnetizing field and an alternating magnetic field with a smoothly decreasing amplitude to zero) at the initial section of the curve does not exceed 1–3% (i.e., less than the error in measuring the parameters of magnetization curve points according to standard methods [2, 6]).

The disadvantage of the technique [27] is the labor-intensive and low accuracy of experimental measurements of the magnetization of the material in a large number of points on the descending and ascending branches of the saturation loop of magnetic hysteresis of the material and the need for statistical approximation of the measurement results for the analytical description of the anhysteretic magnetization curve.

4. Development of an analytical expression to describe the anhysteretic magnetization curve $M(H)$ of a ferromagnetic material

For the analytical description of the anhysteretic magnetization curve of a ferromagnetic material based on the results of H_c, M_r, and M_s measurements of the material, we use that this curve can be calculated as the average line of the descending and ascending branches of the saturation loop of its magnetic hysteresis. The technique [30] for the analytical description of the anhysteretic magnetization curve of a ferromagnetic material is based on the approximation of the change in its magnetization M from the intensity H of the acting magnetizing field on the descending and ascending branches of the saturation hysteresis loop by the Frelich formulae [31–33]:

downstream branch:

$$M = \frac{M_r M_s (H + H_c)}{M_s H_c + M_r H} \quad (1)$$

ascending branch:

$$M = \frac{M_r M_s (H - H_c)}{H_c (M_s - 2M_r) + M_r H} \quad (2)$$

This is justified by numerous experiments and practical uses of formula (Eq. (1)) to approximate the change in the magnetization M of a ferromagnetic material in the second quadrant of the (M, H) plane, which is important for the purpose of the chapter.

For example, R.I. Janus in Ref. [33] on the basis of formula (Eq. (1)) obtained a formula for calculating the remanent magnetization M_d of a ferromagnetic body with demagnetizing factor N after magnetization to saturation:

$$M_d = \frac{M_s(H_c + NM_r)}{2NM_r}\left[1 - \sqrt{1 - \frac{4M_r^2NH_c}{M_s(H_c + NM_r)^2}}\right] \tag{3}$$

In deriving formula (Eq. (3)), R.I. Janus used the well-known [5, 10, 33–35] formula for calculating the magnetic field strength H in a ferromagnetic body in an external magnetic field of strength H_e:

$$H = H_e - NM \tag{4}$$

Experimental verification of formula (Eq. (3)) was carried out in Ref. [36]. The central demagnetization factor N of the used samples varied from 0.00118 to 0.222. The samples had the form of ellipsoids, cylinders, parallelepipeds, and plates made of structural steel and steels 60SiCR7, C 50, and 100Cr6, which underwent different heat treatment. The verification showed almost complete coincidence of the results of calculations and experiments carried out by different authors. No further justification is required for the legitimacy of using formula (Eq. (1)) (and formula (Eq. (2)) – formula (Eq. (1)) shifted by $2H_c$ along the abscissa axis) to achieve the purpose of the chapter.

The values of magnetization M on the descending and ascending branches of the saturation loop of magnetic hysteresis correspond, respectively, to the values of H_1 and H_2 of the magnetizing field strength (**Figure 4**).

From (Eq. (1)) and (Eq. (2)), we obtain:

$$M = \frac{M_rM_s(H_c + H_1)}{M_sH_c + M_rH_1} \tag{5}$$

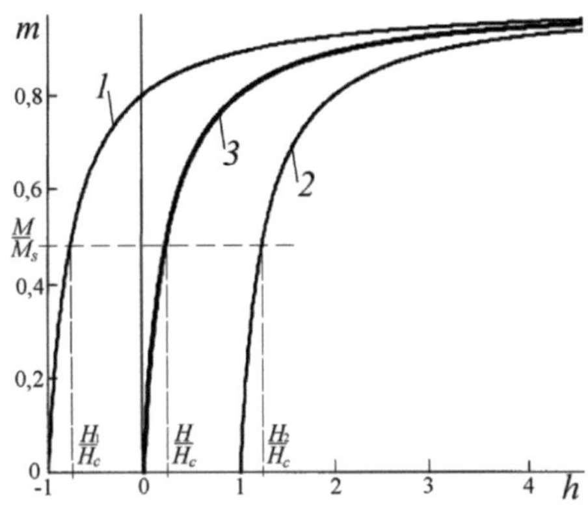

Figure 4.
The descending (1) and ascending (2) branches of the saturation magnetic hysteresis loop and the anhysteretic magnetization curve (3) plotted in dimensionless coordinates ($h = H/H_c$, $m = M/M_s$) using formulae (Eq. (1)), (Eq. (2)) and (Eq. (10)) for a material with $K_r = 0.8$.

$$M = \frac{M_r M_s (H_2 - H_c)}{H_c(M_s - 2M_r) + M_r H_2} \tag{6}$$

Transforming (Eq. (5)) and (Eq. (6)), we find:

$$H_1 = \frac{M_s H_c (M_r - M)}{M_r(M_s - M)} \tag{7}$$

$$H_2 = \frac{H_c(M_s M - 2M_r M + M_s M_r)}{M_r(M_s - M)} \tag{8}$$

The half-sum of the magnetic field strengths H_1 and H_2 (taking into account their sign) is equal to the field strength H on the anhysteretic magnetization curve (**Figure 4**):

$$H = \frac{H_2 - H_1}{2} = \frac{H_c M(M_s - M_r)}{M_r(M_s - M)} \tag{9}$$

from where:

$$M = \frac{M_r M_s H}{H_c(M_s - M_r) + M_r H} \tag{10}$$

For the Stoletov curve of the material at its anhysteretic magnetization from (Eq. (10)):

$$\chi = \frac{M_r M_s}{H_c(M_s - M_r) + M_r H} \tag{11}$$

In the designations: $K_r = M_r/M_s$, $h = H/H_c$, $\xi = H_c/M_s$, let us write (Eq. (11)) in the form:

$$\chi = \left[\xi\left(h + K_r^{-1} - 1\right)\right]^{-1} \tag{12}$$

5. The effectiveness of the use of the developed formula

The effectiveness of using the developed formula (Eq. (10)) to calculate the anhysteretic magnetization curve will be illustrated on the example of comparing the switching and anhysteretic magnetization curves of C 30 steel tempered at different tempering temperatures T_t after quenching from 860°C (**Figure 5**). The results of measurements in Ref. [4] of magnetic parameters H_c, M_r, and M_s of this steel, given in **Table 1**, were used for calculation. **Table 1** also summarizes the experimental values of the maximum χ_m and initial χ_a magnetic susceptibilities of C 30 steel measured in Ref. [4]. The maximum values χ_{mA} of the magnetic susceptibility on the anhysteretic Stoletov curve were calculated using formula (Eq. (13)) derived from formula (Eq. (11)) at H = 0:

$$\chi_{mA} = \frac{M_r M_s}{H_c(M_s - M_r)} \tag{13}$$

The anhysteretic Stoletov curves were constructed using the parameters H_c, M_r, and M_s given in **Table 1** by formula (Eq. (11)), and the commutation curves were constructed using the same parameters by the formulae of reference [9]. In the table, for each heat treatment of C 30 steel, the results of χmA calculation of the maximum

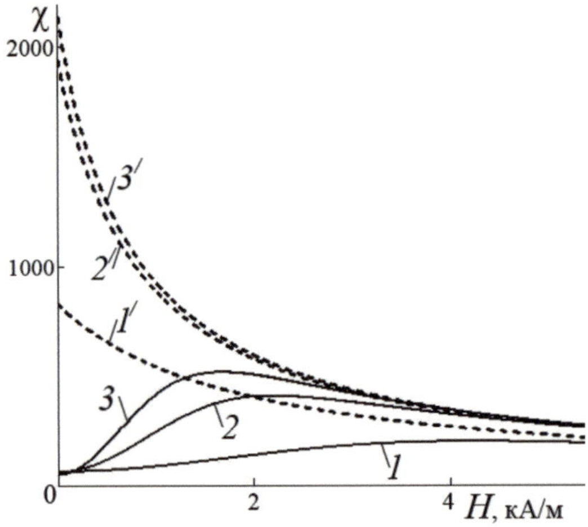

Figure 5.
Normal (1, 2, 3) and anhysteretic (1′, 2′, 3′) Stoletov curves of C 30 steel (respectively samples No. 1, 5, and 9 from ***Table 1****).*

No Sample	T_t, °C	Experiment according to Ref. ([4], Table 1.1)						Calculation			
		HRC	H_c, кА/м	M_r, кА/м	M_s, кА/м	χ_m	χ_a	χ_{mA}	χ_{mA}/χ_m	χ_{mA}/χ_a	N_i •100
1	20	46	2.30	865	1589	254	60	825.4	3.25	13.76	0.1212
2	150	45	2.25	870	1591	245	63	853.2	3.48	13.54	0.1172
3	200	44.5	2.10	876	1599	259	65	922.6	3.56	14.19	0.1084
4	250	43.5	1.43	970	1644	399	74	1654.5	4.15	22.36	0.0604
5	300	44	1.22	1007	1652	489	82	2114.1	4.32	25.78	0.0473
6	350	41	1.13	1070	1650	541	80	2693.8	4.98	33.67	0.0371
7	400	38	0.995	1145	1643	613	77	3796.6	6.19	49.31	0.0263
8	450	34	0.873	1248	1645	699	76	5923.5	8.47	77.94	0.0169
9	500	32	0.876	1265	1648	719	72	6213.6	8.64	86.3	0.0161
10	550	26	0.866	1277	1639	733	69	6676.4	9.11	96.76	0.0150
11	600	23	0.834	1280	1632	761	65	7115.8	9.35	109.5	0.0141
12	650	19	0.730	1235	1622	759	72	7090.6	9.34	98.48	0.0141

Table 1.
Results of measurement and calculation of parameters of C 30 steel.

magnetic susceptibility of steel on the anhysteretic magnetization curve and the results of N_i calculation of these materials using formula (Eq. (16)) are also given. Note that as the strength H of the magnetizing field increases, the anhysteretic magnetization, and switching magnetization curves of the samples (*1* and *1′*, *2* and *2′*, *3* and *3′*) become closer and practically coincide. But when the intensity H of the magnetizing field tends to zero, the difference between the magnetic susceptibilities χ of the material on the anhysteretic and normal magnetization curves of the material increases and reaches a maximum at $H = 0$. At the same value of $H = 0$, and the value

χ_{mA} of the maximum magnetic susceptibility of steel on the anhysteretic magnetization curve reaches a maximum (on the switching magnetization curve at $H = 0$, the magnetic susceptibility has a minimum initial value of χ_a). In this case, the maximum value χ_m of magnetic susceptibility on the normal magnetization curve is observed in the field, the strength of which can be calculated by the formula developed and substantiated in [37]. Note also that the maximum value of χ_{mA} of the magnetic susceptibility of steel on the anhysteretic magnetization curve is simultaneously the value of the initial magnetic susceptibility of steel on this curve.

It is of interest to analyze to what extent this value of χ_{mA} exceeds the values of the maximum χ_m and initial χ_a of the magnetic susceptibilities of C 30 steel measured on the normal magnetization curve. The results of calculation of the ratios χ_{mA} to the experimentally measured in Ref. [4] values of the maximum χ_m and initial χ_a magnetic susceptibilities of this steel measured on the normal magnetization curve are also given in **Table 1**.

Figure 6 shows the dependence between the values of χ_{mA} and χ_m of samples No. 1–12 from **Table 1**. Note that the magnetic parameters H_c, M_r, and M_s of these samples vary within sufficient limits for analysis.

The results presented in **Table 1** show that anhysteretic magnetization increases the maximum magnetic susceptibility of C 30 steel by a factor of 3.25–9.35. In this case (**Figure 6**), the dependence of the maximum magnetic susceptibility χ_{mA} of C 30 steel on the anhysteretic magnetization curve on the maximum magnetic susceptibility χ_m of this steel can be approximated by an exponential dependence in the investigated range of variation of these parameters with high reliability (the reliability of R^2 approximation is 0.9969):

$$\chi_{mA} = 295.82 \cdot e^{0.0042 \cdot \chi_m} \quad (14)$$

The results presented in the table also show that anhysteretic magnetization increases the initial magnetic susceptibility χ_a of C 30 steel by 1.13–2.04 orders of magnitude. But the dependence of χ_{mA} of steel 30 on its χ_a in the investigated range of

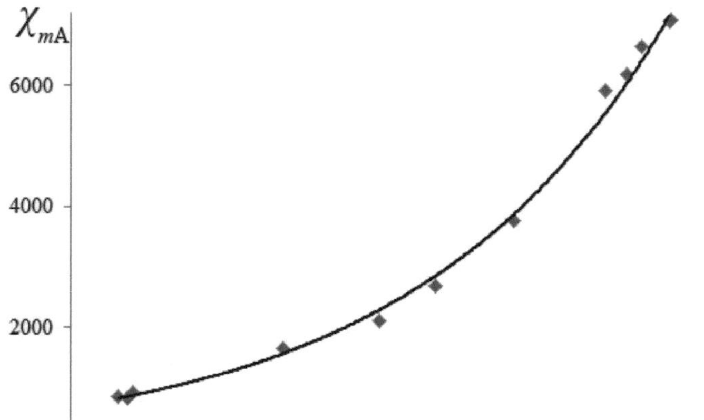

Figure 6.
*Dependence between χ_{mA} and χ_m values of the maximum magnetic susceptibilities of C 30 steel samples (**Table 1**) subjected to different heat treatments, measured on anhysteretic and normal magnetization curves. Exponential trend line of this dependence.*

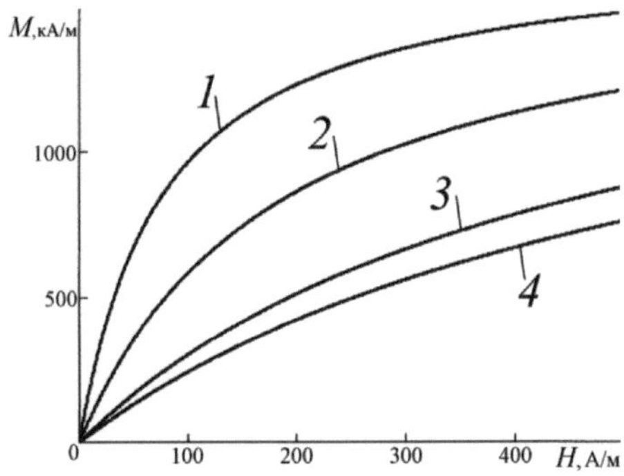

Figure 7.
*Anhysteretic magnetization curves of carbon steels with carbon contents of 0.085, 0.23, 0.32, and 0.52% (curves 1–4, respectively), whose magnetic parameters are given in **Table 2**. Calculation using formula (Eq. (10)).*

variation of these parameters is ambiguous because of the ambiguity of the change of χ_a from the tempering temperature T_t of C 30 steel in this range (**Table 1**). The obtained results of the analysis are new. They could not be obtained based on experimental studies of the anhysteretic magnetization curve of C 30 steel without the developed formula (Eq. (11)). It should be noted that such results can be obtained for any ferromagnetic material by measuring its H_c, M_r, and M_s, which for the majority of known materials are given in scientific and reference literature or can be measured with maximum achievable accuracy by standard methods.

By introducing the notations $K_r = M_r/M_s$, $m = M/M_s$, $h = H/H_c$ and transforming from (Eq. (11)), we obtain:

$$m = \left(1 + \frac{1 - K_r}{K_r h}\right)^{-1} \tag{15}$$

Formula (Eq. (15)) is obtained mathematically accurately on the basis of the experimentally justified in Refs. [10, 27] methodology and experimentally justified in Refs. [33, 36] dependences (Eq. (1)) and (Eq. (2)). It does not require additional experimental substantiation.

The developed technique was used to determine the anhysteretic magnetization curves of carbon steels (**Figure 7**).

The authors of Ref. [3] magnetized samples of the studied steels with carbon content of 0.085, 0.23, 0.32, and 0.52% to saturation in a closed magnetic circuit of the ballistic unit permeameter. Measurements of H_c, M_r, and M_s of the sample material were carried out by ballistic method according to the method [2, 6]. The results of these measurements ([3], **Table 2**) are given in **Table 2**.

Carbon content, %	H_c, А/м	M_r, кА/м	M_s, кА/м
0.085	71	816	1707
0.23	188	836	1659
0.32	399	782	1675
0.52	487	749	1630

Table 2.
Magnetic parameters of carbon steels according to reference [3].

The dependence of magnetization M on the anhysteretic magnetization of carbon steels on the intensity H of the magnetizing field was determined using the results (**Table 2**) of measuring the parameters H_c, M_r, and M_s of the sample material in accordance with expression (Eq. (10)).

6. Calculation and analysis of internal demagnetization coefficient of ferromagnetic materials

Demagnetizing magnetic field—a field inside a magnetized body, the source of which is the magnetization M of the body itself and whose action reduces M. The demagnetization coefficient N is the ratio of the demagnetizing magnetic field in a magnetized body to the magnetization M of this body. In the presence of demagnetizing field, the true (internal) magnetizing field H is determined from the formula (4).

The internal demagnetization N_i is determined by measuring the anhysteretic magnetization curve of thermally demagnetized [1] ferromagnetic material. The slope of the anhysteretic magnetization curve to the M axis is caused by internal fields arising from mechanical stresses, material inhomogeneities, and anisotropy. The tangent of the angle α of this slope when the magnetizing field H tends to zero (**Figure 2**) is equal to N_i. Therefore, N_i of a material allows us to reliably judge its structural parameters and physical and mechanical properties [11, 27, 38]. This is the reason for its widespread use and the importance of its determination for ferromagnetic materials [39–47]. The necessity of carrying out a complete cycle of magnetic effects on thermally demagnetized material, measurements of its magnetization with statistical processing of their results and differentiation of the obtained dependence complicate the process of N_i determination and reduce its accuracy. Meanwhile, it was shown in Refs. [8, 9, 48] that all "specific" changes in magnetic hysteresis loops, magnetic permeability, and other magnetic parameters of steels associated with changes in their structural state and phase composition are caused only by changes in the H_c, M_r, and M_s of these steels. At the same time, H_c, M_r, and M_s of ferromagnetic materials can be measured according to standard methods during switching magnetization of the material with minimal errors compared to other magnetic parameters. The results of H_c, M_r, and M_s measurements of many materials are given in scientific and reference literature.

The purpose of the paragraph is to establish the relation between the internal coefficient N_i of demagnetization of a ferromagnetic material and the parameters H_c, M_r, and M_s of the saturation loop of its magnetic hysteresis, which can be measured for any material by standard methods.

In addition to simplifying the measurement process, the importance of solving the problem at hand is that there is no requirement to perform measurements on thermally demagnetized state material when measuring H_c, M_r, and M_s [1, 6]. Therefore, the results of the effect of thermal treatments on H_c, M_r, and M_s of many materials are given in scientific and reference literature (in contrast to the effect of these treatments on N_i). This difference in the initial requirements to the measured object is the reason for the absence in the (known to the author) scientific and reference literature of the results of measurement of N_i and H_c, M_r, and M_s of the same material. Therefore, it is not possible to compare the results of N_i calculation by the developed formula with the results of experimental N_i measurement. Proceeding from this, in developing the required formula, the author paid primary attention to the substantiation of the physical assumptions of the calculation and mathematical rigor of the transformations carried out. To calculate N_i, we use the developed formula (Eq. (10)) describing the magnetization change on the anhysteretic magnetization curve. Formula (Eq. (10)) was obtained in Ref. [30] mathematically accurately on the basis of the experimentally substantiated methodology [10, 27] and experimentally substantiated in Refs. [33, 36] dependences (Eq. (1)) and (Eq. (2)). It does not require additional experimental substantiation.

By performing the transformations, from (Eq. (10)), we obtain for the internal coefficient N_i of demagnetization:

$$N_i = \frac{H_c(1 - K_r)}{M_r} \tag{16}$$

In Ref. [7], it is recommended in the development of methods and devices of magnetic structuroscopy instead of M_r and H_c parameters to use the results of measurement of ratios $K_r = M_r/M_s$ and $\xi = H_c/M_s$. With this in mind, we get:

$$N_i = \xi \frac{1 - K_r}{K_r} \tag{17}$$

It was shown in Ref. [5] that in the case when the magnetic anisotropy is uniaxial and the direction of the light axis of magnetization varies from section to section, the ratio between M_r and M_s:

$$M_r = 0,5M_s \tag{18}$$

For substances with cubic magnetic anisotropy [5]:

$$M_r = 0,866M_s \tag{19}$$

In ferromagnetic steels M_r has intermediate values with respect to calculation by (Eq. (18)) and (Eq. (19)) [3–5].

Figure 8 shows the change in N_i of the steel when varying the parameter K_r in practically important range $0.5 \le K_r \le 0.866$.

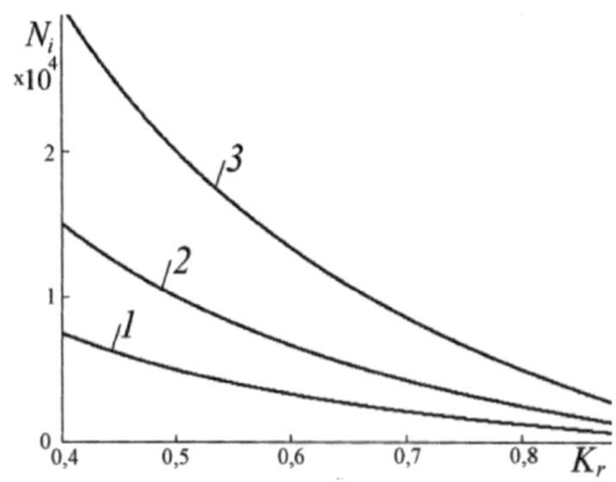

Figure 8.
Effect of changing the parameter K_r (1–3 – when ξ of the material is 0.0005; 0.001 and 0.002, respectively) on the internal demagnetization factor N_i of steel. Calculation by formula (Eq. (17)).

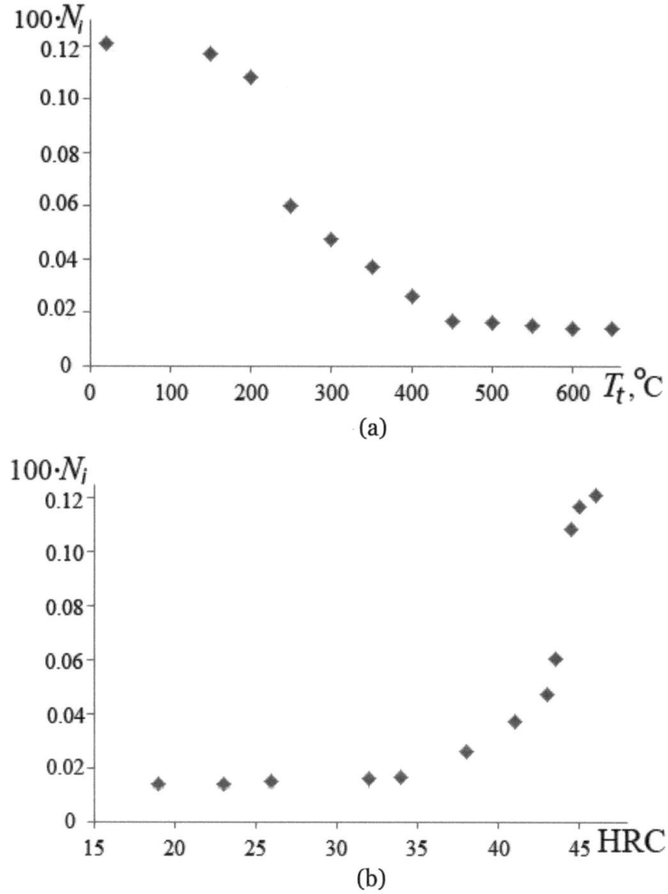

Figure 9.
Effect of tempering temperature T_t (a) and hardness HRC (b) of C 30 steel after quenching from 860°C on its internal N_i demagnetization factor. Calculation by formula (Eq. (16)).

The developed formula (Eq. (16) allows us to analyze the influence of technological modes of obtaining ferromagnetic materials on their internal N_i demagnetization coefficient.

As an example of such an application of the developed formula, **Figure 9** shows the results of the analysis of the change in the internal coefficient N_i of demagnetization of C 30 steel from its tempering temperature T_t. The magnetic parameters of C 30 steel after different tempering temperatures T_t measured in [4] using standard methods are given in **Table 1**.

The obtained result shows the monotonicity of $N_i(T_t)$ and N_i(HRC) dependences in the whole range of T_t variation (in contrast to the non-monotonicity of this dependence for the H_c parameter in the range 450°C $\leq T_t \leq$ 550°C). This result was obtained on the basis of measurements by standard methods of H_c, M_r, and M_s parameters of C 30 steel, given in the reference book [4]. Experimental construction of the dependences obtained in **Figure 9** using the standard N_i determination technique is impossible, since fulfillment of the requirement [1] to construct an anhysteretic magnetization curve on thermally demagnetized material (i.e., after heating it above the Curie temperature) will distort the effect of the tempering temperature T_t of the sample on its magnetic properties.

7. Determination of the anhysteretic magnetization curve of a ferromagnetic body from its demagnetizing factor and magnetic parameters of the material

Magnetic elements used in electrical engineering are often ferromagnetic bodies (plates, solid or hollow cylinders, and rings) whose material magnetization is influenced by their demagnetization factor *N* [35]. This determines the importance of determining and analyzing the anhysteretic magnetization curve of a ferromagnetic body. It is also important for analyzing the anhysteretic magnetization of dispersed ferromagnetic particles. Their shape can be considered as ellipsoids of rotation, which are magnetized uniformly.

The aim of the paragraph is the analytical description of the anhysteretic magnetization curve of a ferromagnetic body.

Calculation of the anhysteretic magnetization curve of a ferromagnetic body.

From (Eq. (10)) and (Eq. (4)):

$$M = \frac{M_r M_s (H_e - NM)}{H_c(M_s - M_r) + M_r(H_e - NM)} \tag{20}$$

Transforming (Eq. (20)), we obtain the reduced quadratic equation for determining *M*:

$$M^2 - \left[\frac{H_c(M_s - M_r)}{M_r N} + \frac{H_e}{N} + M_s\right] M + \frac{M_s H_e}{N} = 0 \tag{21}$$

Equation (Eq. (21)) has the following solution:

$$M = \frac{1}{2}\left[\frac{H_c(M_s - M_r)}{M_r N} + \frac{H_e}{N} + M_s\right] - \sqrt{\frac{1}{4}\left[\frac{H_c(M_s - M_r)}{M_r N} + \frac{H_e}{N} + M_s\right]^2 - \frac{M_s H_e}{N}}. \tag{22}$$

Formula (Eq. (21)) for describing the magnetization change on the anhysteretic magnetization curve of a ferromagnetic body is obtained mathematically accurately on the basis of the physically reliable relationship (Eq. (4)) and the dependence (Eq. (10)) of the magnetization change on the anhysteretic magnetization curve of a ferromagnetic material substantiated in [30]. Therefore, it does not require additional experimental substantiation.

7.1 Analysis of the influence of body sizes and magnetic properties of the material on their anhysteretic magnetization curves

As an example of calculation by formula (Eq. (22)), **Figure 10** shows anhysteretic magnetization curves of cylindrical rods of different relative length λ made of classical electrical materials: technically pure iron (M_s = 1715,7 kA/m, M_r = 940 kA/m, H_c = 80 A/m) and textured siliceous iron (M_s = 1560 kA/м, M_r = 798 kA/m, H_c = 8 A/m) ([34], Tables 12.4, 12.5). The demagnetizing factors *N* (central demagnetization coefficients) of the rods were calculated by the formula of K. Warmuth [35, 49] for $\lambda > 1$:

$$N(\lambda) = \frac{1}{\lambda^2 - 1}\left[\frac{\lambda}{\sqrt{\lambda^2 - 1}} \ln\left(\lambda + \sqrt{\lambda^2 - 1}\right) - 1\right] \frac{1 + 2,35\ln(1 + 0,137\lambda)}{1 + 2,28\ln(1 + 0,284\lambda)}. \quad (23)$$

For a generalized analysis of the influence of body sizes and magnetic properties of the material on their anhysteretic magnetization curves, we use the following notations:

$$K_r = \frac{M_r}{M_s}, \xi = \frac{H_c}{M_s}, m = \frac{M}{M_s}, h_e = \frac{H_e}{H_c}. \quad (24)$$

Carrying out transformations, from (Eq. (22)) taking into account (Eq. (24)), we obtain:

$$m = F(h_e)\left[1 - \sqrt{1 - \frac{h_e \xi}{N[F(h_e)]^2}}\right] \quad (25)$$

where:

$$F(h_e) = \frac{\xi}{2N}\left[\frac{1 - K_r}{K_r} + h_e + \frac{N}{\xi}\right] \quad (26)$$

Note (curves *1*, *1′* in **Figure 10**) that as the demagnetizing factor N of the body increases, the magnetic properties of its material cease to influence the anhysteretic magnetization curve in the practically important range of its variation. We also note that the parameters N and ξ are included in expressions (Eq. (25)) and (Eq. (26)) for the anhysteretic magnetization curve of a ferromagnetic body in the form of the ratio ξ/N. Taking this into account, introducing the parameter γ:

$$\gamma = \frac{\xi}{N} \quad (27)$$

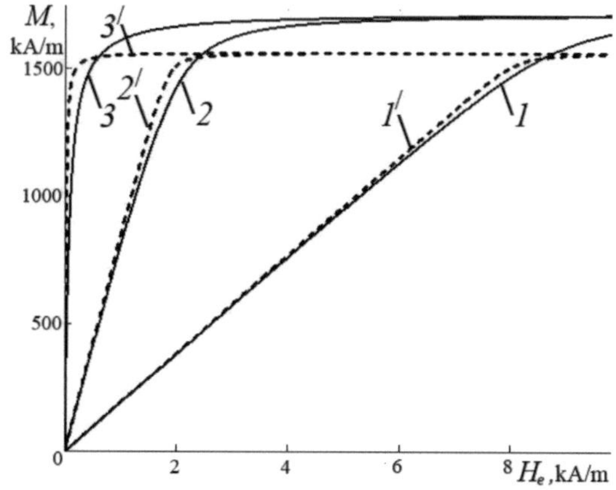

Figure 10.
Anhysteretic magnetization curves M(H$_e$) of cylindrical rods with length-to-diameter ratio λ of 20 (1, 1′), 50 (2, 2′) and 1000 (3, 3′) made of technically pure iron (1, 2, 3) and textured siliceous iron (1′, 2′, 3′). Grows according to (Eq. (22)) and (Eq. (23)).

transform (Eq. (25)) and (Eq. (26)) to a convenient form for generalized analysis:

$$m = F(h_e)\left[1 - \sqrt{1 - \frac{h_e\gamma}{[F(h_e)]^2}}\right] \quad (28)$$

where:

$$F(h_e) = \frac{\gamma}{2}\left[\frac{1-K_r}{K_r} + h_e + \frac{1}{\gamma}\right] \quad (29)$$

It follows from formulas (Eq. (28)), (Eq. (29)) that a decrease in the demagnetizing factor N of the body changes its anhysteretic magnetization curve $m(h_e)$ (according to the change in its parameter γ) at a constant ratio $\xi = H_c/M_s$ of the material as much as an increase in the ratio ξ of the material at a constant N of the body.

From formulas (Eq. (28)), (Eq. (29)), it also follows that the rectangularity factor of the magnetic hysteresis loop of a material affects the change of the anhysteretic magnetization curve of a ferromagnetic body by the change of the ratio $(1–K_r)/K_r$.

The obtained results of the analysis of the influence of the magnetic properties of the material and demagnetizing factor of ferromagnetic bodies on the character of changes in their anhysteretic magnetization curves could not be obtained without using the developed formulas (Eq. (22)), (Eq. (28)), and (Eq. (29)) as a result of the analysis of experimentally measured anhysteretic magnetization curves of ferromagnetic bodies or their numerical calculations by the finite element method.

8. Conclusions

Thus, as a result of the research conducted:

1. Formula $M = \frac{M_r M_s H}{H_c(M_s - M_r) + M_r H}$ (Eq. (10)) for the analytical description of the anhysteretic magnetization curve of a ferromagnetic material and formula $\chi = \frac{M_r M_s}{H_c(M_s - M_r) + M_r H}$ (Eq. (11)) for technical calculations of the Stoletov $\chi(H)$ curve under anhysteretic magnetization of a material have been developed and substantiated. The developed formulas use the results of measuring the saturation magnetization M_s, remanent magnetization M_r, and coercive force H_c of the material. Increased productivity and simplified determination of the anhysteretic magnetization curve and Stoletov curve of the material is achieved due to the fact that for their determination, it is necessary to magnetize the material only once to technical saturation and to measure after that only three magnetic parameters of the material's saturation hysteresis loop. The used parameters H_c, M_r, and M_s can be measured according to standard methods with minimal error compared to other magnetic parameters of the material. They are given in Ref. and scientific literature for many steels, cast irons, and alloys. This makes it possible to replace labor-consuming and inaccurate measurements of the anhysteretic magnetization curve and Stoletov curve of materials under anhysteretic magnetization by calculation using the developed formulas.

2. Using the developed formula (Eq. (11)), the values of initial and maximum magnetic susceptibilities of C 30 steel after different heat treatments under anhysteretic and normal magnetization curve were compared. This allowed us to quantitatively establish that anhysteretic magnetization increases the initial magnetic susceptibility of C 30 steel by 1.13–2.04 orders of magnitude and increases its maximum magnetic susceptibility by 3.25–9.35 times.

3. Based on the reliable physical assumptions and the developed formula (Eq. (10)), formula $N_i = \frac{H_c(1-K_r)}{M_r}$ (Eq. (16)) for determining the intrinsic coefficient N_i of demagnetization of a ferromagnetic material using the measurement results of its H_c, M_r and M_s is mathematically accurately developed. Of the parameters that can be determined by a nondestructive method, N_i of ferromagnetic materials by its physical nature is the material parameter most sensitive to changes in its structure. Formula (Eq. (16)) is intended to analyze the influence of chemical composition and heat treatments of ferromagnetic materials (steels, cast irons, and porous materials) on their N_i (i.e., on their structure).

4. On the basis of reliable physical assumptions and the formula (Eq. (10)), a formula $M = \frac{1}{2}\left[\frac{H_c(M_s-M_r)}{M_rN} + \frac{H_e}{N} + M_s\right] - \sqrt{\frac{1}{4}\left[\frac{H_c(M_s-M_r)}{M_rN} + \frac{H_e}{N} + M_s\right]^2 - \frac{M_sH_e}{N}}$

(Eq. (22)) was mathematically accurately developed for the analytical description of the anhysteretic magnetization curve of a ferromagnetic body based on the results of measurement or calculation of its demagnetization coefficient N and the main magnetic parameters of the material: H_c, M_r, and M_s. The combinations of ferromagnetic body parameters depending on its N and magnetic properties that determine the magnetization change on the anhysteretic body magnetization curve have been established. The obtained results of the analysis of the influence of these parameters on the anhysteretic magnetization curves of a body could not be obtained by analyzing the experimentally measured anhysteretic magnetization curves of ferromagnetic bodies. They are important for understanding the process of anhysteretic magnetization of ferromagnetic bodies and its effective use.

Author details

Sergey G. Sandomirski
Joint Institute of Mechanical Engineering, NAS of Belarus, Minsk, Belarus

*Address all correspondence to: sand_work@mail.ru

References

[1] GOST 19693 – 74. Magnetic materials. In: Terms and Definitions. Moscow: Standards Publishers; 1974. 32 p. (in Russian)

[2] GOST 8.377–80. Magnetically soft materials. In: Methodology of Measurements for Determination of Static Magnetic Characteristics. Moscow: Standards Publishers; 1986. 22 p. (in Russian)

[3] Miheev MN, Morozova VM. Magnetic and Electrical Properties of Steel after Different Types of Heat Treatment. Moscow: ONTI CNIIKA; 1964. 46 p. (In Russian)

[4] Bida GV, Nichipuruk ΛP. Magnctic Properties of Heat-Treated Steels. Ekaterinburg: UrO RAN; 2005. 218 p. (In Russian)

[5] Tikadzumi S. Physics of Ferromagnetism. New York: Oxford University Press; 1997. 672 p. ISBN-10: 0-19-851776-9. ISBN-13: 978-0-19-851776-4

[6] Chernyshev ET, Chechurina EN, Chernysheva NG, Studencov NV. Magnetic Measurements. Moscow: Standards Publishers; 1969. 248 p. (in Russian)

[7] Sandomirski SG. Enhancement of structural sensitivity of remanent magnetization and coercivity of steels. Defektoskopija [Russian Journal of Nondestructive Testing]. 2023;**8**:62-64. (in Russian)

[8] Sandomirski SG. Calculation of the main magnetization curve of structural steels based on parameter measurements of the saturation hysteresis loop. Measurement Techniques. 2017; **60**(2):177-182. DOI: 10.1007/s11018-017-1170-y

[9] Sandomirski SG. Calculation of the Stoletov curve of structural steels from the parameters of the saturation hysteresis loop. Elektrichestvo [Electricity]. 2022;**1**:18-23. DOI: 10.24160/0013-5380-2022-1-18-23. (in Russian)

[10] Bozorth RM. Ferromagnetism. New York: D Van Nostrand Company, Inc.; 1951. 867 p

[11] Polivanov KM. Ferromagnetics. In: Fundamentals of the Theory of Practical Applications. Moscow–Leningrad: State Encyclopedic Edition; 1957. 256 p. (in Russian)

[12] Janus RI. Magnetization curves. In: Physical Encyclopedic Dictionary. Vol. 3. Moscow: Soviet Encyclopedia; 1963. pp. 354-355. (in Russian)

[13] Silveyra J.M., Conde Garrido J.M. On the anhysteretic magnetization of soft magnetic materials, AIP Advances, 2022. Vol. 12. 3. id. 035019, 13 p. Available from: https://aip.scitation.org/doi/10.1063/9.0000328

[14] Szewczyk R. Validation of the anhysteretic magnetization model for soft magnetic materials with perpendicular anisotropy. Materials. 2014;**7**:5109-5116. DOI: 10.3390/ma7075109

[15] Szewczyk R. Assessment of uncertainty of determination of parameters of Jiles-Atherton model of hysteresis loops of isotropic materials, Przegląd Elektrotechniczny (Polish) [Electrotechnical Review] 1(2016), p. 164–165. 10.15199/48.2016.11.40

[16] Nowicki M. Anhysteretic magnetization measurement methods

for soft magnetic materials. Materials. 2018;**11**:2021. DOI: 10.3390/ma11102021

[17] Pearson J, Squire PT, Atkinson D. Which anhysteretic magnetization curve? IEEE Transactions on Magnetics. 1997;**33**:3970-3972. DOI: 10.1109/20.619632

[18] Silveyra JM, Conde Garrido JM. On the modeling of the anhysteretic magnetization of homogeneous soft magnetic materials. Journal of Magnetism and Magnetic Materials. 2021;**540**:168430. DOI: 10.1016/j.jmmm.2021.168430

[19] Volchkov SO, Duhan AE, Duhan EI, Kurlyandskaya GV. Automated computer-aided inspection Center for Magnetoimpedance Spectroscopy. Russian Journal of Nondestructive Testing. 2016;**52**(11):647-652

[20] Shevcov DA, Shishov DM, Troshin PA, Ivanov NS. Meazuring complex for registration of hysteresis loops of ferromagnetic materials. Russian Electrical Engineering. 2022; **93**(1):11-15

[21] Walker J, Saddow SE. Noncontact Anhysteric Curve Plotter and Stats Feld to Time-Varying Hysteressgraph with Integrated Temperature Chamber. Pat. US 8,274.278 B1; Published 25.09.2012. Patent Official Gazettes OG20120925_1382-4

[22] Rao DK, Kuptsov V. Effective use of magnetization data in the design of electric machines with overfluxed regions. IEEE Transactions on Magnetics. 2015;**51**:6100709. DOI: 10.1109/TMAG.2015.2397398

[23] Jiles DC, Atherton D. Ferromagnetic hysteresis. IEEE Transactions on Magnetics. 1983;**19**:2183-2185. DOI: 10.1109/TMAG.1983.1062594

[24] Kokornaczyk E, Gutowski MW. Anhysteretic functions for the Jiles–Atherton model. IEEE Transactions on Magnetics. 2015;**51**:7300305. DOI: 10.1109/TMAG.2014.2354315

[25] Jiles DC, Kiarie W. An integrated model of magnetic hysteresis, the magnetomechanical effect, and the Barkhausen effect. IEEE Transactions on Magnetics. 2020;**57**:800211. DOI: 10.1109/TMAG.2020.3034208

[26] Liu L. How the B-H Curve Affects a Magnetic Analysis (and How to Improve it). COMSOL, Inc.; 2019. Available from: https://www.comsol.com/blogs/how-the-b-h-curve-affects-a-magnetic-analysis-and-how-to-improve-it/ [Accessed: September 16, 2024]

[27] Stashkov AN, Somova VM, Sazhina EY, Stashkova LA, Nichipuruk AP. Magnetic method for determining the concentration of residual austenite in maraging steels. Russian Journal of Nondestructive Testing. 2011;**47**(12):810-814

[28] Yamazaki K, Sato Y, Domenjoud M, Daniel L. Iron loss analysis of permanent-magnet machines by considering hysteresis loops affected by multi-axial stress. IEEE Transactions on Magnetics. 2019;**56**:7503004. DOI: 10.1109/TMAG.2019.2950727

[29] Upadhaya B, Rasilo P, Perkkiö L, Handgruber P, Belahcen A, Arkkio A. A constraint-based optimization technique for estimating physical parameters of Jiles–Atherton hysteresis model. COMPEL-The International Journal for Computation and Mathematics in Electrical and Electronic Engineering. 2020;**39**:1281-1298. DOI: 10.1108/COMPEL-08-2019-0332

[30] Sandomirski SG. Determining the anhysteretic magnetization curve for a

ferromagnetic material according to the limiting loop parameters of its magnetic hysteresis. Russian Electrical Engineering. 2023;**94**(10):753-757. DOI: 10.3103/S1068371223100103

[31] Frolich O. Üder des Magnetisierungsgestz des Eisens [About the magnetization law of iron]. Elektrotechnische Zeitschrift. H. 1894; **27**:368-370. (in German)

[32] Frolich O. Üder die magnetischen Konstanten des weichen Eisens [about the magnetic constants of soft iron]. Elektrotechnische Zeitschrift. H. 1894; **38**:517-519. (in German)

[33] Janus RI. Magnetic Defectoscopy. Gostehizdat: Moscow–Leningrad; 1946. 171 p. (in Russian)

[34] Jiles D. Introduction to Magnetism and Magnetic Materials. London: Chapman & Hall; 1989. 440 p

[35] Sandomirski SG. Calculation and Analysis of Demagnetizing Factor of Ferromagnetic Bodies. Minsk: Belarusian science; 2015. 244 p. ISBN 978-985-08-1862-1. (in Russian)

[36] Sandomirski SG. Sensitivity of remanent magnetization of ferromagnetic components to the magnetic characteristics of their materials and geometrical parameters. Soviet Journal of Nondestructive Testing. 1991;**26**(12):875-881

[37] Sandomirski SG. Estimated calculation of the magnetic field at which the magnetic permeability of steels reaches a maximum. Elektrichestvo [Electricity]. 2012;**7**:63-68. (in Russian)

[38] Gorkunov ES, Zakharov VA, Ul'yanov AI, Chulkina AA. Influence of pore shape and orientation on the internal demagnetization coefficient of porous magnetic materials. Defektoskopija [Russian Journal of Nondestructive Testing]. 2001;**3**:31-37. (in Russian)

[39] Gorkunov ES, Zakharov VA, Chulkina AA, Ul'yanov AI. Internal demagnetization factor for porous ferromagnets in remagnetization process. Russian Journal of Nondestructive Testing. 2004;**40**(1):1-7

[40] Sandomirski SG. Estimation of the internal demagnetizing factor of cast iron from its measured remanent magnetization. Russian Metallurgy (Metally). 2013;**5**:392-397

[41] Anhalta M, Weidenfeller B, Mattei JL. Inner demagnetizing factor in polymer bonded soft magnetic composites. Journal of Magnetism and Magnetic Materials. 2008;**320**:e844. DOI: 10.1016/j.jmmm.2008.04.061

[42] Skomski R, Hadjipanayis GC, Sellmyer DJ. Effective demagnetizing factors of complicated particle mixtures. IEEE Transactions on Magnetics. 2007; **43**(6):2956-2958. Available from: https://digitalcommons.unl.edu/mrsecfacpubs/71

[43] Takács J, Kovács G, Varga LK. Internal demagnetizing factor in ferrous metals. Journal of Metallurgy. 2012;**2012**: 752871. DOI: 10.1155/2012/752871. 5 pages

[44] Varga LK, Franco V, Kákay A, Kováčb J, Mazaleyrat F. The role of internal and external demagnetizing effects in nanocrystalline alloys. Transactions on Magnetics. 2001;**37**(4): 2229-2231

[45] Varga LK, Kováčb J, Novák L. Determination of external and internal demagnetizing factors for strip-like amorphous ribbon samples. Journal of

Magnetism and Magnetic Material. 2020;
507:1-5

[46] Onderko F, Bircáková Z, Kollár P, Füzer J, Strecková M, Szabó J, et al. Influence of ferrite and resin content on inner demagnetizing fields of Fe-based composite materials with ferrite-resin insulation. Acta Physica Polonica A. 2020;**137**(5):846-848. DOI: 10.12693/APhysPolA.137.846

[47] McCann SM, Leach J, Reddy SM, Mercer T. Methods of investigating the demagnetization factors within assemblies of superparamagnetic nanoparticles. AIP Advances. 2022;**12**: 075212-1-075212-12. DOI: 10.1063/5.0095899

[48] Sandomirski SG. Calculation of magnetization variation on partial hysteresis loops of steels from the main magnetic parameters of the saturation hysteresis loop. Elektrichestvo [Electricity]. 2016;**12**:39-43. (in Russian)

[49] Warmuth K. Über den ballistischen Entmagnetisierungsfaktor zylindrischen Stäbe [About the ballistisc demagnetization factor of cylindrical rods]. Archiv für Elektrotechnik. 1954; **41**(5):242-257. (in German)

Chapter 6

PVDF Based Flexible Ferroelectrics

Mudassar Shehzad, Liu Yang and Yaojin Wang

Abstract

The well-known ferroelectric polymer, Poly (vinylidene fluoride-trifluoroethylene) [P(VDF-TrFE)], has a larger polarization-electric field hysteresis loop and possesses a high degree of crystallinity than poly(vinylidene fluoride-trifluoroethylene-clorotrifluoroethylene) [P(VDF-TrFE-CTFE)]. [P(VDF-TrFE-CTFE)] is a relaxer that contains its third monomer, CTFE, and breaks the coherence of its polarization domains in a nano polar region. It eventually shows narrower polarization PE loop characteristics than a normal ferroelectric polymer. The blended solution was dispensed in freestanding films and its beta phase enhancement was further studied which may have an impact on its electroactive characteristics. The electroactive properties, especially pyroelectric (PyE), piezoelectric (PE), and flexoelectric (FLE) properties of the blends are associated with its beta phase characteristics. Flexible electronics benefit greatly from the use of electroactive devices; however, a significant drawback of this technology is its higher power consumption combined with somewhat lower efficiency. To address this issue, we fabricated thin film devices designed to transmit and record electrical signals for pyroelectric and piezoelectric transduction within a single unit. These devices were constructed using a straightforward technique that involved creating sandwiched layered structures with a small radius. Ultimately, the printed thin film devices, which contained the enhanced ß phase, functioned effectively as both pyroelectric energy harvesters and piezoelectric transducers, among other applications.

Keywords: Olsen cycle, ferroelectric polymer, energy harvesting, electrostriction, polarization, antiferroelectric (AFE) loop, energy density, piezoelectric properties, sound pressure level, sound intensity, and piezoelectric transducers

1. Introduction

Polyvinylidene fluoride (PVDF) is recognized as an electroactive polymer that is used in both industrial and medical purposes. However, its widespread adoption is hindered by its lower density and reduced electric strain capabilities. A major challenge in utilizing these polymers is finding ways to decrease their high-power density through various technological approaches. For instance, in 2011, researchers led by Zhang discovered a method to enhance the electrostrictive performance of copolymers and terpolymers, which allows for greater energy density in PVDF. Nevertheless, this advancement involves intricate synthesis techniques and complex polymerization processes [1]. Recently, P(VDF-TrFE) copolymer has shown a high electric polarization and electromechanical (EM) response compared to other established ferroelectric materials [2]. Researchers are currently concentrating on PVDF-based copolymers and terpolymers due to their remarkable ferroelectric characteristics. The copolymer shows substantial electric strain when subjected to high-energy electron irradiation of the films. These irradiated films display high energy density, making them highly appealing for a variety of electromechanical applications [3–5]. **Figure 1** illustrates three configurations of polymer chains: 'trans-gauche-trans-gauche' (TG + TG-), all 'trans' (TTT), and long trans 'TTTG+TTTG-'.

In the polar region of P(VDF), the packing morphology contributes to the enhancement of its polarization. The α phase, typically associated with TG+ and TG- conformations, lacks polarization due to the antiparallel alignment of the polymer chains in this amorphous form. In contrast, the β phase, where two all-trans chains are oriented parallel to each other, is crystalline and exhibits highly notable electrical properties due to its substantial dipole moment within each unit cell. The polar γ phase is characterized by alternate TTTG+ and TTTG- conformations, with dipole moments aligned in the same direction. After irradiation, PVDF polymers are converted into relaxor ferroelectrics, achieving high dielectric and polarization constants. Introducing structural defects is the primary mechanism by which conventional ferroelectrics are transformed into relaxor ferroelectric chains. Furthermore, relaxor

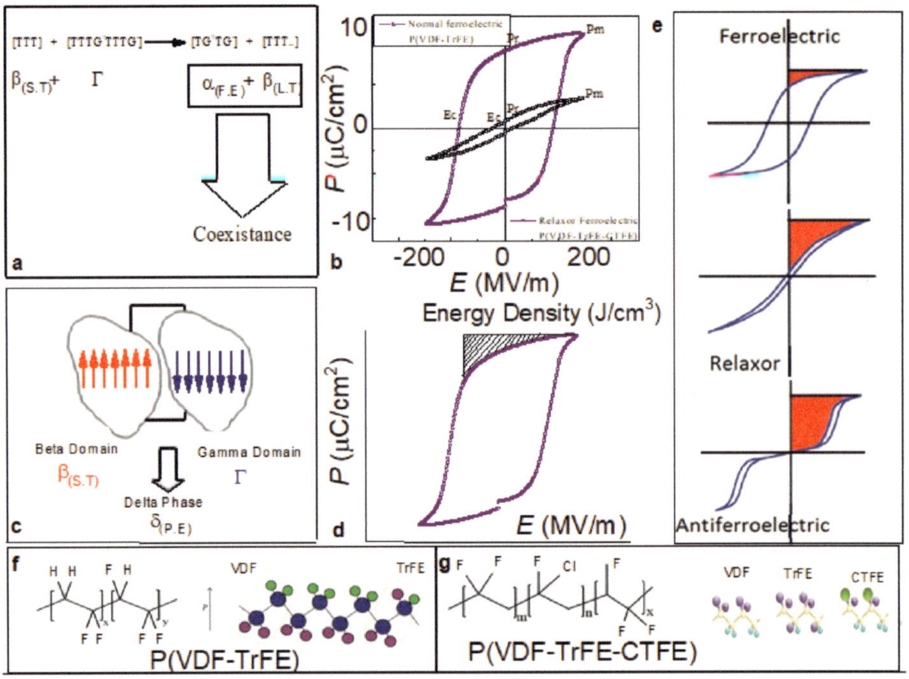

Figure 1.
AFE property of polymer. (a) The three (3) conformations such as long (trans) and short (trans) and trans gauche are responsible for the ferroelectric behavior, where α phase is regarded as the Ferroic Alpha phase. (b) Shows the conversion of co to the terpolymer after complete incorporation of (CTFE) into the PVDF-TrFE copolymer. (c) Shows an interaction among ß and Γ phases which results delta phase. (d) Shows the measurement of energy density (stored). (e) The graphical representation of ferroelectric (FE), Releaxor Ferroelectric (RFE) and Antiferroelectric (AFE) hysteresis loops. (f, g) The molecular chains of PVDF-TrFE and P(VDF-TrFE-CTFE) polymers.

ferroelectric properties can also be attained by substituting a third CTFE monomer within the polymer chain.

Antiferroelectric (AFE) behavior is distinguished by double hysteresis loops (DHL), which feature a central notch. This distinctive, propeller-shaped double hysteresis loop is a unique characteristic of AFE polymers, contributing to their high dielectric constants and substantial energy density. Among the different phases, the β phase stands out as the most important, as it offers strong crystalline properties essential for AFE double hysteresis loop behavior. Traditional ferroelectric materials have the lowest energy density, exhibiting broad polarization, whereas ferroelectric polymers generally achieve high maximum polarization (Pm) and low remnant polarization (Pr). Polymers can be classified into ferroelectric, paraelectric (PE), and antiferroelectric (AFE) types. Paraelectric and antiferroelectric polymers, often referred to as linear materials, have tunable polarization that varies with the applied electric field. For linear materials, energy density is directly proportional to the square of the electric field [6].

Upon applying an electric field, the dipoles align with the direction of the applied field (E). Notably, in materials with a high-energy dielectric density, the dipole orientation (DO) shifts, causing the electric field to either become trapped or released. Antiferroelectric polymers, with an almost zero net dipole moment, are the only means of generating stored electricity in this way. Thus, the antiferroelectric-like (AFE) behavior of the displacement electric (DE) field loop is essential, as it ensures full energy release during electric field conversion. The charged and discharged energy densities are assessed from the DE loop, and calculated by approximating the area under the shaded region in the DE hysteresis loop. The unreleased (or undischarged) energy density is defined as $U_{undischarge} = 100^{*}(1\text{-}U_{released}/U_{stored})$, where U_{stored} has the same value as $U_{unreleased}$ [6].

Flexible piezoelectric materials are recognized to change mechanical energy into electrical energy (direct piezoelectric materials). It can also change the electrical energy to mechanical energy (inverse piezoelectric materials) [1–3]. The inverse and direct piezoelectric materials can detect and transmit electric signals proficiently. For the flexible electronic devices and those having complex forms and manuring may need novel materials that can be stretchy and flexible to perform the essential work. In this persistence, a piezoelectric coefficient plays a huge role in choosing the right material, such as, the piezoelectric coefficient of ferroelectric ceramics and single crystals is in order of 1000 and 4000 pC/N respectively [4, 5]. The piezoelectric coefficient of PVDF polymers is much smaller than that of piezoelectric ceramics [7]. The main disadvantage of perovskite ceramics is that they are massive, heavyweight, and stiff, and cannot be fit to use in a flexible and wearable device. The stiffened nature of these ceramics makes it inappropriate to use them in various applications. To improve the piezoelectric features of polar polymer, the polymer composites can be used that can enhance the piezoelectric coefficient and preserve its elasticity too [8–14]. The incompatible electrical and mechanical features are a key task when adding any kind of ceramic filler to the polar polymer matrix. It may delay the performance of the stretchy piezoelectric polymer composites. The mismatch in dielectric permittivity between ceramic and polymer introduces a discontinuity in electric polarization at the matrix-filler interface, potentially forming an electric double layer that impedes the dielectric response of ceramic fillers. This disparity in permittivity across the matrix and filler can result in polarization loss, ultimately affecting performance. Various methods are being employed to establish conductive pathways for both electric polarization and mechanical load within the polymer matrix to enhance the strength of polymer composites. For instance, polymers are reinforced in three dimensions using a low magnetic field [15], using freezing of ice to build complex composites [16], or bio-inspired assembly of nanoplatelets in a polymer matrix [17, 18]. However, these strategies generally result in a high-volume fraction of ceramic fillers, significantly compromising the mechanical flexibility of ceramic-polymer composites [6, 8, 19–21].

The key objective of our research work is to produce and select the right piezoelectric material for flexible acoustic devices like microphones and magnet-free loudspeakers. Since realizing the piezoelectric features of PVDF-based polymers, scientists have been looking for ways to fit in transparent piezoelectric devices into today's rapidly developing smart device ethos [21]. PVDF polymers are lighter in weight, transparent, and can be processed by using simple techniques [22–25]. PVDF polymers have good ferroelectric, pyroelectric, piezoelectric, and flexoelectric properties, making them perfect for transparent actuators and sensors. Investigators also tried to study the electroacoustic (EA) features of polymers. One such application that utilizes its inverse piezoelectric effect is the transparent thin film loudspeaker. Unlike traditional magnetic speakers, thin film loudspeakers operate primarily at a higher voltage. The high operating voltages of such piezoelectric materials limit their widespread use in many current technologies such as low-power mobile screens or smartphones etc. However, traditional speakers also have great limitations in their applications such as heavy and their reliability on electrical coils.

The main characteristic of ferroelectric polymers is the presence of switchable polarization and its domain boundary structure. Ferroelectric and resistive memory, electromechanical, and transport features are related to polarization switching behavior [26]. PVDF-based ferroelectric polymers have fantastic energy harvesting, electrostatic, and conductivity and excellent sensing properties in comparison to other ferroelectric polymers [27, 28]. Recent studies have shown that ferroelectric blends are able to control the structural properties of these materials and have good results in increasing phase percentage [29]. Gomes et al. stated that the degree of stretch and temperature greatly influences the transition from the non-polar alpha phase to the polar beta phase in PVDF-based polymers. Recently, Alexander De. Neef et al. studied that PVDF mixed with PMMA [30] has a piezoelectric coefficient (d_{33}) of up to 11 pCN^{-1}. Therefore, by poling the PVDF blends, the beta phase percentage, piezoelectric properties, and electrochemical properties can be improved. Chen et al. studied the acoustic properties of P (VDF-TrFE) and P (VDF-TrFE-CTFE) compounds by blending precise amount of copolymer in the beta phase of terpolymers [31]. The non-polar alpha phase progresses gradually and slowly to the polar beta phase. The copolymer is mixed in proportion to the terpolymer which enhances the beta phase reflection because of electrical and thermal poling.

Here we propose that we can enhance the pyroelectric, piezoelectric, and flexoelectric coefficients of the blends. The novel fabrication technique is adopted to enhance the electroactive properties of the polymer blends. It will also enhance the piezoelectric and flexoelectric coefficient with the improvement in stiffness and by imparting the stretchability in the blend matrix. The chain structure of the corresponding polymer phase is given in **Table 1**.

1.1 Material preparation

Copolymer, P(VDF-TrFE) 70/30 mole% is normal FE, and terpolymer, P(VDF-TrFE-CTFE) 70/20/10 mole% is a Relaxor FE. These polymers are subjected to characterization for their inherent properties, after mixing with each other in a definite proportion. Terpolymer and copolymer were dissolved in methyl isobutyl ketone (MIBK) separately

Phases	Dipole alignment	Polarization	Comments	Chain Structure
Alpha (α)	↑↓	✗	TG^+, TG^-	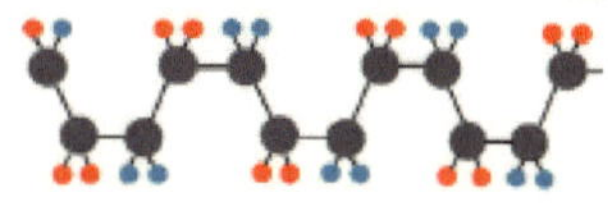
Beta (ß)	↑↑	✓	TTTT	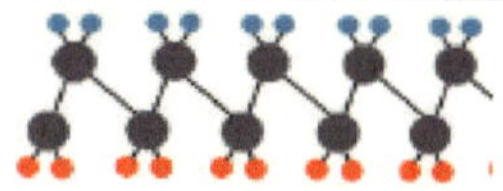
Gamma (Γ)	↓↓	✓	$TTTG^+$, $TTTG^-$	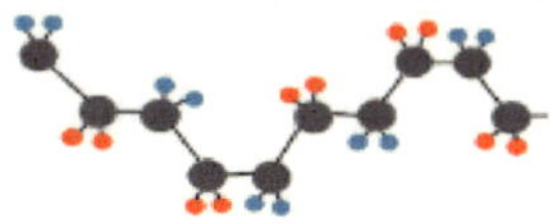
Delta (δ)	↑↓	✗	Paraelectric Phase	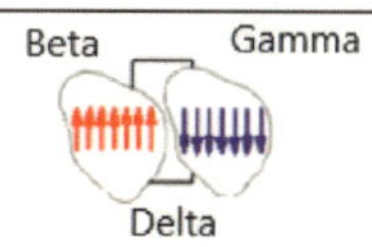
• Hydrogen • Fluorine ● Carbon				

Table 1.
Phases and their polarization responses based on literature review [32].

and mixed with each other in a definite stoichiometric ratio. Blend solution of 10% of P(VDF-TrFE) and 90% of P(VDF-TrFE-CTFE) is called here as α = 1 where α is copolymer content in a definite proportion of terpolymer solution. Also, α = 0 is 0% copolymer in 100% terpolymer, 0.1 is 10% copolymer in 90% of terpolymer solution, and so on. The film was fabricated using blends ranging from α = 0 to 1. Consequently, the films thickness ca. 20 to 30 microns were fabricated by the hand and machine layup method by adopting the solution cast method. The blend is dispensed at the glass plate to fabricate the thin films, and these films were cast by the machine-layup method to maintain its smoothness. Additionally, the cast films were annealed in heating furnaces to improve their crystalline properties. We have also observed that the annealing temperature governs as a key factor in predicting various properties of the above solution 0 < =α < =1.

2. Energy harvesting for flexible electronics

2.1 Pyroelectric energy harvesting (PyEH)

Energy storage is needed in the modern era, especially after the industrial revolution in this world. Scientists are trying to construct self-powered and independent devices with less power consumption. The demand for these devices is growing day after day in electronics industries [33]. Consequently, scavenging methods are developed by scientists in these methods wasteful energy is stored [34–37]. The major types of scavenging energy harvesting techniques include wasteful energy harvesting from mechanical energy by using the piezoelectric properties, thermal energy harvesting through the Olsen power cycle using the pyroelectric effect [38], magnetic energy harvesting and photoelectric energy harvesting through solar cells [39].

Figure 2 represents the schematics of ferroelectric polymer heating at high temperature, because of the Olsen power cycle that ultimately harvests the energy due to the additional polarization by charge buildup at electrodes by reducing the temperature. It also indicates the alignment of domains upon heating the pyroelectric material up to the ferroelectric to paraelectric phase transition temperature.

In Pyroelectric energy harvesting, the wasteful heat is difficult to capture and labeled as a low-grade waste product [41]. To capture the wasteful scavenging heat energy, thermoelectric and/or pyroelectric energy harvesting practices can be used. If the antiferroelectric energy harvesting is carried out at ambient temperature; the technique is defined as room temperature energy harvesting technique, while if the energy storage takes place by tuning the temperature of antiferroelectric polymers; referred to as pyroelectric energy harvesting [42]. The pyroelectric material uses the wasteful heat energy and converts it to the electrical energy using the temporal temperature gradient [43, 44].

The antiferroelectric polymers store the room temperature electrical energy if these are made up as capacitors [45]. Instead, the pyroelectric polymers are used as a thermal sensing device. If the energy storage capacity of these materials is high enough to charge the low voltage electronic devices, then these materials can be used as pyroelectric energy harvesters shown in **Figure 3**. Since the 1960s the discovery of the pyroelectric effect, it is still considered as a less studied area [46–48].

2.2 Wasteful heat energy harvesting

Interestingly, the high ECE of antiferroelectric materials has attracted great interest due to their superior cooling efficiency compared to conventional methods such as gas evaporation [49, 50]. ECE is the adiabatic temperature difference of a dielectric material as a result of the dipolar effect generated during the action of an

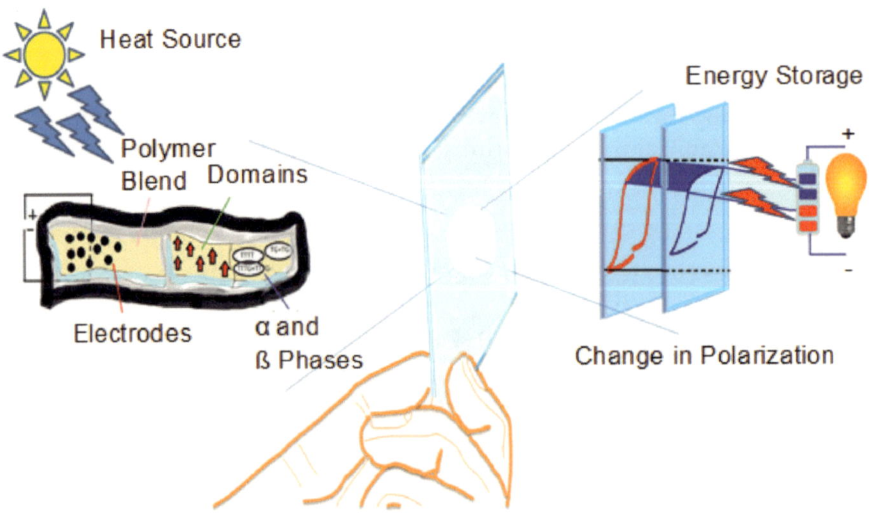

Figure 2.
Schematics of ferroelectric polymer heating at high temperature (i.e., 25°C < T < Tc) where Tc is the curie temperature [40].

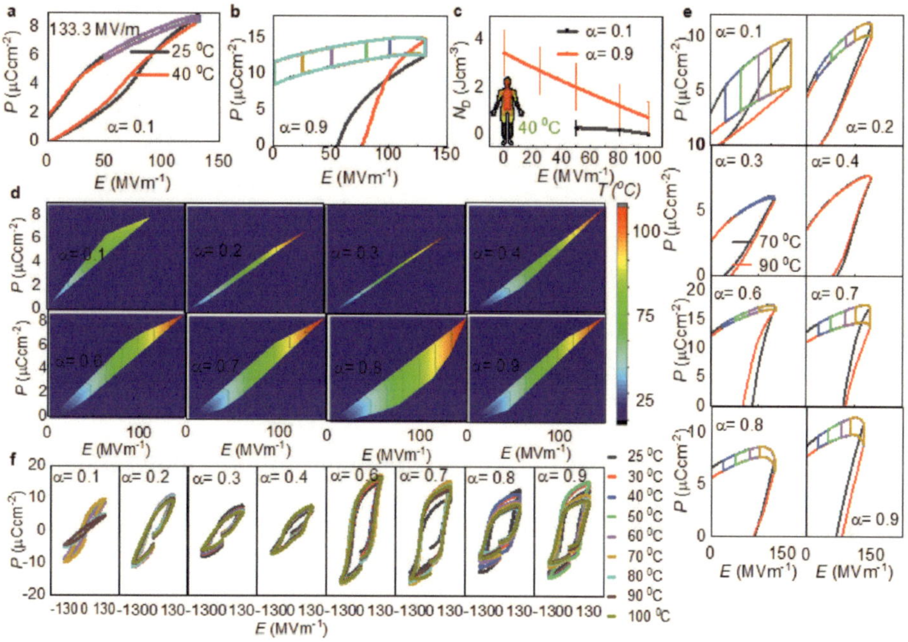

Figure 3.
The pyroelectric energy harvesting (PyEH) near a body temperature at T_1 = 25°C (77°F) and T_2 = 40°C (104°F) at E_H = 133 MV/m at 100 Hetrz frequency. (a) α = 0.1, (b) α = 0.9, and (c) shows the comparison of energy density (N_D) for α = 0.1 and 0.9 blends. (d) Shows the polarization-electric field and temperature contour graph for α =0.1, 0.2, 0.3, 0.4, ... 0.9 blends. (e) Shows Olsen power cycle for α = 0.1 to α = 0.9 blends EL = 0 MV/m EH = 133.33 MV/m from 70 to 90°C at 100 Hz. (f) The temperature-dependent P-E hysteresis loop at E = 133.33 MV/m at 100 Hz frequency of α = 0.1 to 0.9 blends. (Standard deviation: n = 8) [40].

electric field [51]. It is well-known that the dipoles of a dielectric material follow an alignment path applied to an electric field. Changing the direction of the electric field will then reverse the orientation of this dipole [10]. The variation of the direction of the electric field causes a certain degree of randomness, also called the entropy of the system. Such an ordered dipole scheme leads to smoother effects of the dielectric material. Various types of materials have been reported to exhibit ECE effects, such as Polyvinylidene difluoride (PVDF) and so on.

Ferroelectric polymers gained the remarkable consideration of scientists' community because of their valuable large ECE [10, 51], enhanced dielectric constants [52, 53], improved electric properties (i.e., energy density), and remarkably improved electro-mechanical-actuation [54]. Electrocaloric effects can also be attributed to heat transfer from the PVDF samples to the ferroelectric heat source and to the paraelectric heat transfer coefficient. The δ-phase paraelectric phase at high temperatures, while the ß-phase plays a role in the crystallization phase of the material at high temperatures, which is due to the presence of the bulk TrFE group in the bulk chain of the PVDF-based polymer [5]. Here, the Beta (ß) phase is under consideration where it experiences a phase transition between the normal FE phase and the piezoelectric (PE) phase [14]. **Figure 4** shows the adiabatic temperature difference, which depends on the frequency acting on the constant electric field (E (MV/m)). In the ß phase, the polymer chains are usually aligned in the TTTT configuration, and the chain alignment is dependent on the dipole orientation. The temperature changes are responsible for the phase changes of all transitions to TG, such as TGTG′ and TTTG′ [3].

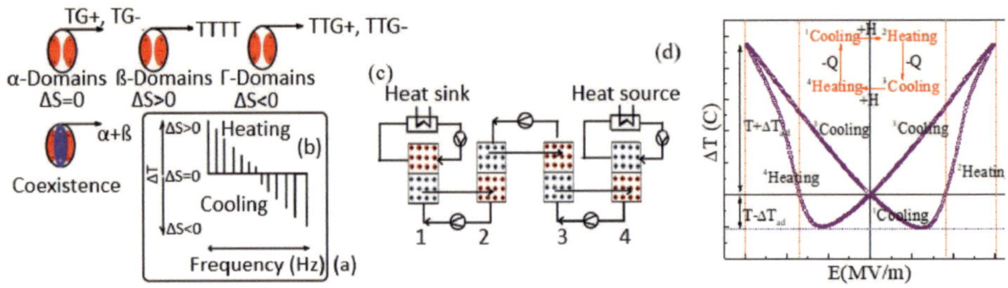

Figure 4.
Schematics of dipolar array. (a) shows randomness degree which helps to change its phase from TTTT to TGTG'. More randomness will be achieved if reduce the frequency, (b) gives the information of variation in adiabatic temperature (ΔT_{ad}) with decrease in frequency, (c) represents the illustration of heating cycle of a FE based Olsen heat engine and (d) shows the ΔTE loop [55].

The change in temperature (variation in adiabatic temperature) addressed and reported as 10°C when electric field is reduced to zero [3]. According to the ECE Landau–Devonshire theory, the entropy change in the isotherm is directly proportional to the square of the electrical conductivity with respect to ΔD.

$$\left[\Delta S = \frac{-1}{2}\beta_{LD}\Delta D^2\right] \tag{1}$$

where, $ß_{LD}$ is the term used for Landau-Devonshire coefficient from his theory [15].

P(VDF-TrFE-CTFE) and P(VDF-TrFE) are two types of polymers that have been extensively studied due to their ECE effects. The electric field in a dielectric material is defined as $D = E_0E + P$, where D is the electrical conductivity of the material, P is the polarization ($\mu C/m^2$) and E_0 is the free space permittivity. It is generally agreed that there are two types of dielectric polymers, linear and nonlinear polymers. In polymers, P_0 increases the strength of the electric field called linear material. During the phase change from ferroelectric to paraelectric, the temperature gradient at the Curie temperature (Tc) is determined by the thermal conductivity of the material. Various scientific publications have well-documented the differences between P(VDF-TrFE) and P(VDF-TrFE-CTFE) as secondary to F-P (ferroelectric paraelectric) is the polarization-dependent free energy of the polymer second-order phase transition, and it can be written in Eq. 2,

$$G = G_0 + \frac{1}{2}\beta_{LD}(T - T_c)P^2 + \frac{1}{4}\xi P^4 - EP \tag{2}$$

Eq. 2 β_{LD} can be reached, where T is the temperature determined by the DE path temperature (298.15) K, Tc = Currie temperature (°C or K), $\mathcal{E} = \mathcal{E}_0\mathcal{E}_r$ = dielectric permittivity, C/Vm, Polarization = (C/m^2) and E = electric field (10^6 V/m). The adiabatic change in temperature (ΔT) and entropy change (ΔS) are directly related to mechanical stress and strain, the rate is always linear.

$$\Delta T = \frac{-1}{2C_E}\beta_{LD}TD^2 = \frac{T\Delta S}{C_E} \tag{3}$$

ΔS (Joule/Kg*K), which represents the response to entropy change, and ΔT (°C) represents the adiabatic temperature response due to dipole disorientation. Cp is

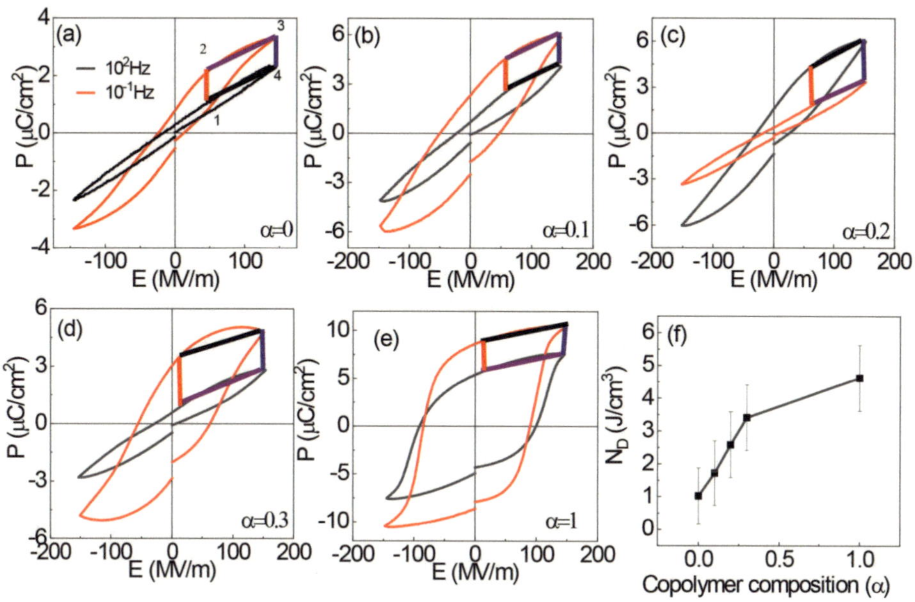

Figure 5.
Frequency dependent Olsen power cycle. a–e The Olsen power cycle of blends at 10^2 and 10^{-2} Hz. f Energy density comparison of polymer blends depending on the copolymer contents [55].

the specific heat capacity, β is the thermal conductivity ((Joule*m)/(K*C²)), and C_E (Joule/Kg*K) is the thermal conductivity of the material used in the DSC values. More specifically, the specific heat capacity Cp (Joule/K) is expressed as,

$$\frac{dH}{dt} = C_P \frac{dT}{dt} + f(t,T) \tag{4}$$

Here, dH/dT = Heat Flow Signal (mWatt), dT / dt = Heating rate (10°C/min)) and m = mass (mg). The change in the enthalpy (ΔH) measured by area under the curve i.e.

$$\Delta H = \int_{T_1}^{T_2} C_E dT = \frac{Joule}{Kg * K} * K = \frac{Joule}{Kg} or \frac{Joule}{g} \tag{5}$$

Figure 5 shows the voltage-dependent Olsen force cycle. **Figure 5(a–e)** shows the integrated Olsen power cycles at 10^2 and 10^{-2} Hz. **Figure 5(f)** shows the energy balance of the polymer blends based on the copolymer core.

The ferroelectric properties, dielectric properties and electrocaloric susceptibility of P(VDF-TrFE), P(VDF-TrFE-CTFE) were investigated herein. Using mixing concentrations, the field and frequency dependence of ECE on blends are investigated.

3. Flexible and transparent piezoelectric transducers

3.1 Flexible and transparent piezoelectric loudspeaker

These days the need for lightweight and thin piezoelectric loudspeakers is rising for mobile phones and other touchscreen applications. Nowadays, researchers are fabricating low-powered thin, and transparent loudspeakers. Various types of loudspeakers exist such as coil loudspeakers, electrostatic loudspeakers, piezoelectric loudspeakers, etc. The high-quality reproduction of sound is obtained by the traditional coil loudspeakers with permanent magnets to generate sounds [56].

These coil loudspeakers are bigger in volume and heavier in weight therefore cannot be used in smart devices [57]. While the electrostatic loudspeaker (ESL) uses the electrostatic field-activated thin diaphragm with low distortion it lacks the base response during its performance [58].

The piezoelectric loudspeakers were announced to overcome the disadvantages of coil and ESL loudspeakers [59]. It can overcome the high voltage problem of ESL because it uses an electric field to convert it into sound without using any magnetic coil. The piezoelectric loudspeakers use the electric field to produce the sound by altering its dimensions in response to the input signals. The mechanical actuation of the diaphragm ultimately alters its electric field into the sound. The frequency response in piezoelectric loudspeakers is limited and needs to be enhanced. If it is improved then it can transform the touch screen industry due to its exclusive feature of miniaturization, flexibility, and quick response for the loudspeaker business. The rewards of piezoelectric loudspeakers cover better transient response without moving coil, lower damping loss, no interference of external magnetic field, high impedance and low harmonic distortion etc. [60]. **Figure 6** represents the schematics of transparent and flexible piezoelectric loudspeakers created by electroactive polymer blends.

In 2003 Li et al. proposed the flexible loudspeaker which has the sound pressure level 80 dBSPL [62]. But in 2009 Sugimoto et al. presented his PVDF-based flexible and transparent loudspeaker with better performance [63]. In this research, we have proposed a loudspeaker with 91 dBSPL and better noise reduction characteristics. We have reported that the transduction property of piezoelectric loudspeakers is directly linked with their beta phase characteristics shown in **Figure 7**. Similarly, we have developed the flexoelectric microphone based on the flexoelectric effect discussed in Section 3.2 in detail. The high sensitivity is attained due to the zeta phase which is due to the reaction between α and ß phases shown in FTIR spectra.

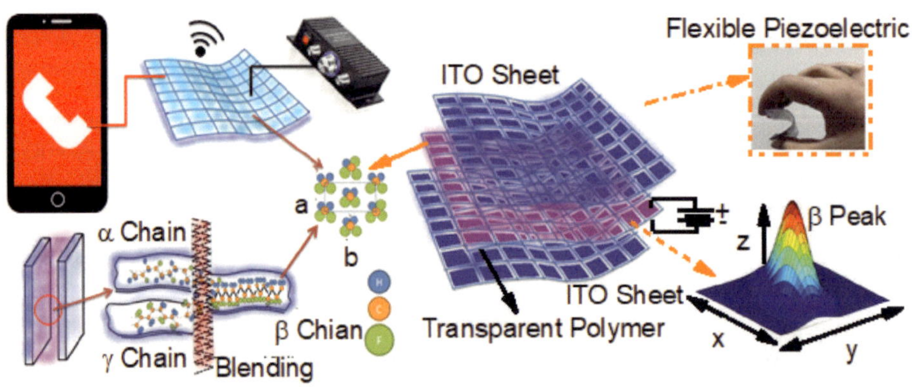

Figure 6.
The schematics of a flexible, transparent, and durable piezoelectric speaker, which is fabricated by wrapping the polymers between a pair of two (2) indium-titanium oxide electrodes. The schematics also represent the ß chain structure of copolymer which is fabricated in the result of the reaction between their α and Γ phases during blending [61].

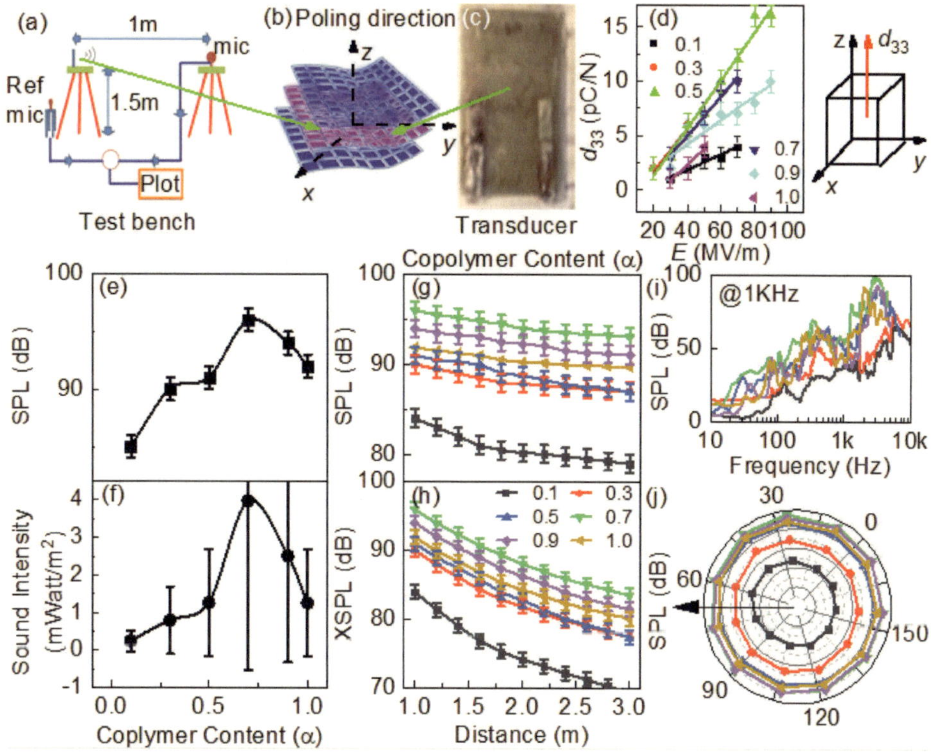

Figure 7.
Shows the transduction and piezoelectric properties of polymer speakers. (a) The schematic diagram of the test configuration for measurement of acoustic properties according to ISO standard. (b) Shows the poling direction (c) Picture of an actual device. (d) d_{33} of blends. (e) Sound Pressure Level graph at one meter away from the target at c.a 10^2 Vpp. (f) Describes the SI (i.e., W/m²) plot at one meter away from the target at 10^2 Vpp (approx.). (g) Shows the SPL Vs distance from the target (m) graph. (h) The inverse graph at various distances (m). (i) Gives the information on sound pressure level vs. frequency at 1 kilohertz. (j) Shows the color-polar trend of sound pressure level at 1 kilohertz. (Standard deviation: n = 6) [61].

3.2 Flexible and transparent flexoelectric microphone

The polarization behavior of PVDF-based polymers is influenced by their polar and non-polar phases, specifically their flexoelectric properties. A critical factor in enhancing polarization is the packing morphology of the polar β-phase. Fourier transform infrared spectroscopy analysis aids in elucidating the trans gauche' configuration, commonly referred to as the α (amorphous) phase. In this α phase, the polymer chains are packed in an antiparallel manner, resulting in a lack of polarization or polarity. Conversely, in the ß phase, the two TTTT domain configurations align in parallel orientations, contributing to the material's polarization capabilities [8]. The electrical properties of the beta phase (crystal) are particularly intriguing, primarily due to the inherent dipole moment found in its lattice cell. In the polar gamma phase, dipole moments are aligned in the same direction, resulting from the alternating conformations of TTTG+TTTG−. Following irradiation, PVDF-based polymers can be converted into relaxor ferroelectrics, which exhibit large polarization and high dielectric constants. This transformation from standard ferroelectrics to relaxed ferroelectric chains is mainly driven by the formation of structural defects. Additionally, incorporating a different CTFE monomer in the polymer chain can also lead to the development of relaxor ferroelectric properties [20, 21].

This study involved the integration of a standard ferroelectric with a relaxor ferroelectric, which enabled us to observe a seamless transition between polar and non-polar phases. Consequently, drawing from the previous literature, we believe that the reflection peak in the FTIR spectrum contributes to the enhanced FE and flexoelectric (FLE) characteristics of the blend system which appears at 1337 cm^{-1} of FTIR spectra (see **Table 2**). In our previous research, we fabricated transparent and flexible piezo-speaker (TFPS) [22]. In this study, we fabricated a transparent and flexible flexo-microphone (TFFM) made of a P(VDF-TrFE) blend, as shown in **Figure 8**. **Figure 8(a)** The schematic illustrates a TFFM integrated into a smartphone screen. The transparent microphone utilizes flexoelectric properties to facilitate the conversion of acoustic signals into electrical signals. This process generates an electrical potential onto the transparent electrode, which is eventually used to recharge the internal battery. The perforation for acoustic transduction is created at the top of the TFFM, as depicted in the illustration **Figure 8(b)**. The TFFM transducer is produced using a straightforward method that involves placing a polymer mixture between two transparent electrodes. When acoustic vibrations occur, the flexible microphone device becomes activated, causing it to oscillate and generate a strong polarization (P). **Figure 8(c)** illustrates the schematic of a flexoelectric device that oscillates to generate sharp polarization. The mechanical drive of this electrical bending device is facilitated by the stress field (ε). In **Figure 8(d)**, a schematic representation of copolymer blending within a terpolymer matrix is displayed, resulting in a well-mixed chain structure of the copolymer and terpolymer molecules. **Figure 8(e)** presents a real photograph of the TFFM sensor. The relationship between its FLE effect, P, and mechanical stress is depicted in **Figure 8(f)**. Thanks to the increment of the zeta (z) phase, we report a FLE coefficient of $\mu = 23\ \mu C\ m^{-1}$ and a sensitivity of $S = 15\ V\ Pa^{-1}$, or 0.4 pC Pa^{-1}, for $\alpha = 0.5$ laminates at a frequency range of $10^1 \leq \omega\ (Hz) \leq 10^2$ at 10^2 dB SPL.

Experimental wavenumber (cm^{-1})	Vibration/group	Comments
3430	Symmetric stretching/N-H	—
3016	Symmetric stretching/CH2	—
2971	Asymmetric stretching/C-H	—
1453	In-plane bending or scissoring/CH2	—
1335 ± 2	In-plane bending or (wagging or twisting/ CH2-CF2	Zeta = α + ß (This Work)
840	CH2 Rocking and CF2 Asymmetric Stretching/ CH2/CF2	γ phase
763	In-plane bending or Rocking/CH2/CH2	α phase
745	In-plane bending or Rocking/CH2/CF2	ß phase
615	CF2 bending and CCC skeletal vibration/CH2/ CCC	α phase
500–1000	Bending/C-F	HH TT configuration
510	Bending and Wagging/CF2	ß phase
490	Bending and Wagging/CF2	α phase
445		ß phase

Table 2.
FTIR spectra for P(VDF) based (co/ter) polymers and its blends [61].

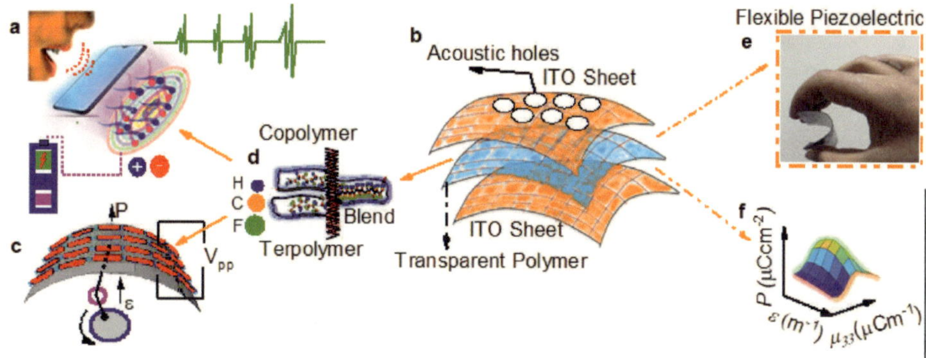

Figure 8.
The schematic diagram illustrates the TFFM microphone. (a) This diagram depicts a TFFM integrated into a mobile screen, capable of sensing and storing FLE energy to charge the smart device. (b) It displays the demonstration of piezoelectric laminates situated between two transparent electrodes, forming the TFFM transducers. (c) The schematic illustrates the ferroelectric (FE) effect of the piezoelectric transducer. (d) This part demonstrates the mixing of co and ter-polymer laminates. (e) A real-time photograph of the TFFM transducer is presented here. (f) The graphical representation outlines the mechanism of FE energy harvesting [61].

These values represent the highest recorded in the literature for the 70/30 P(VDF-TrFE) laminate, which previously had μ_{33} = 0.2μC m^{-1} and S = 10.9 V Pa^{-1} from 50 to 70 Hz.

Figure 9 illustrates the TFFM laminate created using blends ranging from 0 to 1 ($0 \leq \alpha \leq 1$). The schematic diagram of the transparent FE microphone laminate is shown in **Figure 9(a)**. **Figure 9(b)** presents an image of the actual microphone laminate fabricated in the lab. The transparency percentages of the transducers are depicted in **Figure 9(c)**, where the blends with α = 0.3 and α = 0.9 exhibit higher transparency compared to the other blends.

Piezoresponse Force Microscopy (PFM) images for blends with α = 0.1, 0.3, 0.5, and 1.0 are provided in **Figure 9(d)**. **Figure 9(e)** presents the piezoelectric coefficient behavior (d33 in pC/N) of the laminates, revealing that the d33 values for α = 0.5 and α = 0.7 are the highest. Additionally, **Figure 9(f)** details the FLE properties of the transparent microphone laminate, indicating that the α = 0.5 blend demonstrates greater ferroelectric properties than the other polymer blends.

A comparison of the FE coefficient (μ_{33}) across the blend system is illustrated in **Figure 9(g)**. The sensitivity of the laminates at various sound pressure levels is shown in **Figure 9(h–n)**, where the α = 0.5 blend exhibits the highest sensitivity among the tested blends. Consequently, we conducted real-time device tests on the α = 0.5 blend, with the results plotted in **Figure 9(o)**, demonstrating the microphone transducer's response to heart signals. Finally, **Figure 9(p)** displays the sound intensity (SI) as indicated by an oscilloscope for laminates ranging from α = 0 to 1, confirming that the sound intensity for α = 0.5 is greater than that of the other blend systems.

3.3 Piezo-capacitive two-in-one flexible transducer

In touch screen applications for smart devices, condenser transducers (CTs), also called piezo-capacitive transducers, have become essential. Unlike piezoresistive transducers, which detect shifts in impedance, CTs respond to variations in capacitance [64, 65]. A limitation of this technology is its need for a higher electric field to consistently achieve significant polarization (P) In PVDF-based condenser transducers, achieving significant polarization can require nearly hundreds of megavolts

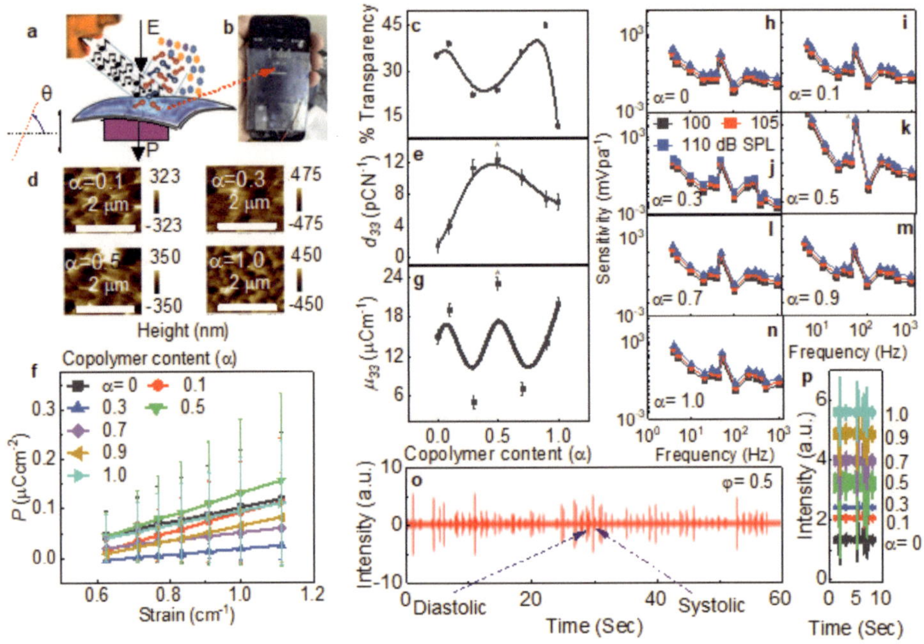

Figure 9.
Shows the TFFM laminate. (a) Depiction of FLE transducer laminate. (b) Photographs of lab-based laminates. (c) transparent (%) of laminates. (d)Piezoresponse Force Microscopic (PFM) photographs of blends. (e) Piezoelectric coefficient, d_{33} (pC/N) behavior of transducers. (f) FLE characteristics of transducers. (g) μ_{33} graphs of blends. (h-n) Sensitivity graphs at different SPL. (o) Real-time response of 50% (α = 0.5) transducer to check the actual heartbeat. (p) SI was checked by oscilloscope for various transducers (SD: n = 7) [61].

(around 100 MV/m). Recently, researchers have been focused on developing low-powered, flexible, and transparent transducer loudspeakers [66]. Consequently, recent literature reflects efforts to develop these transducers with reduced power consumption [60, 67]. Similarly, CTs utilize PVDF-based ferroelectric polymers also require high power to achieve optimal acoustic performance.

Currently, researchers are working to create low-power microphones tailored for potential smart device applications [68–70]. Modern smart societies require active machine/human interaction through smart household electronics applications. To facilitate this interaction, researchers are currently focusing on developing biomimetic multiresonant piezoelectric acoustic sensors (MRPEAS) [71–75] to improve the fidelity of CTs for total cochlear implants to enhance its at high electric field [76–78]. Recently, our group developed a TFPS transducer that achieves an enhanced sound pressure level (SPL) of 91 dB. We fabricated the laminates without insulating-paper material, which helped maintain their transparency [61].

Figure 10 Schematic diagrams of condenser transducers (CTs) include the following: on the left, a schematic representation of the Two-In-One (TIO) transducers, featuring the Condenser Microphone Unit (CMU); on the right, a schematic illustration of the Condenser Loudspeaker Unit (CLU).

CTs (i.e., CLU and CMU) are utilized to enhance fidelity, efficiency, and durability by incorporating suitable circuitry that compensates for their inherent properties. These CTs can function in either manner: as an input/output to an amplifier [80]. As a result, polymeric films have been engineered with embedded microstructures to enhance their acoustic properties, improving fidelity for voice recognition and recording applications

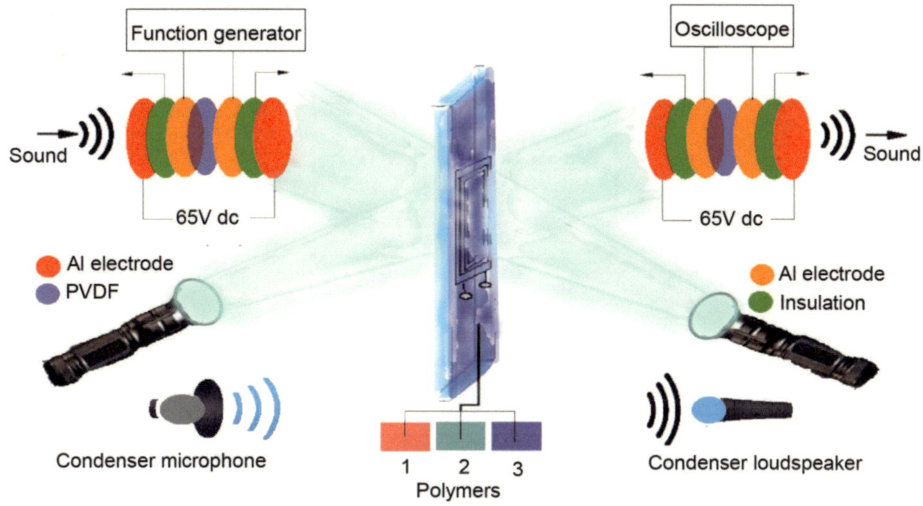

Figure 10.
Schematics of CTs [79].

[81]. The design and area of the active CTs are both crucial for achieving higher sound pressure levels (SPL) and optimal sound recognition. To address the challenges associated with larger area form factors, researchers have fabricated a diverse range of polymer film sensors, with results documented in recent reviews [82–86]. These designs enable the laminates to bend for enhanced performance efficiency; however, during bending, the layers become restricted, which ultimately reduces their acoustic performance [87].

Recently, a miniature condenser transducer (CT) featuring a silicon nitride membrane and a silicon backplate was developed for recording airborne sound. By simply stretching the diaphragm, the device achieved a sensitivity of approximately 1 V/Pa for sound signals ranging from −60 to −90 decibels (dB) [88]. Another transducer was reported that utilizes a silicon nitride diaphragm measuring 1.5 × 1.5 mm, along with a backplate. This design achieved a flat frequency response ranging from 10^2 Hertz to 10^1 kilohertz and a sensitivity of 0.19 mV/Pa with a bias voltage of approximately 65 V [89] fabricated using the sacrificial layer technique, a new Microelectromechanical Systems (MEMS) condenser transducer (CT) was developed featuring a porous diaphragm. By applying spin-on-glass (SOG) as a sacrificial electrode, this design achieved c.a. ~3 pF with a pull-in voltage of 1 megavolt.

Figure 11 shows the TIO CTs of blends. **Figure 11(a)** shows the SPL of blends at V = 0 and 65 V_{dc}. **Figure 11(b)** shows the temperature-dependent SPL (dB) which shows that the SPL is also dependent on the temperature. Also, **Figure 11(c)** provides information on the sound pressure level (SPL) and its associated noise across various frequencies. However, **Figure 11(d)** shows the polar graph of its noise at different frequency range. **Figure 11(e,f)** shows the square law graphs of blends at V = 0 and 65 V_{dc}. Consequently, **Figure 11(g)** shows the frequency visualization at V = 0 and 65 V_{dc} of polymer blends (Alpha, Beta, Gamma, Delta, and Exi) respectively.

This study presents the development of the TIO, CLU and CMU using a simple technique that sandwiches a PVDF laminate between a pair of two Al electrodes, each with a diameter of 3 cm^2. We achieved a sound pressure level (SPL) of 98 dB for the CLU at a capacitance of 0.51 nF for a Gamma blend with a voltage of 65 Vdc. The condenser unit demonstrated remarkable speech recognition capabilities, achieving a

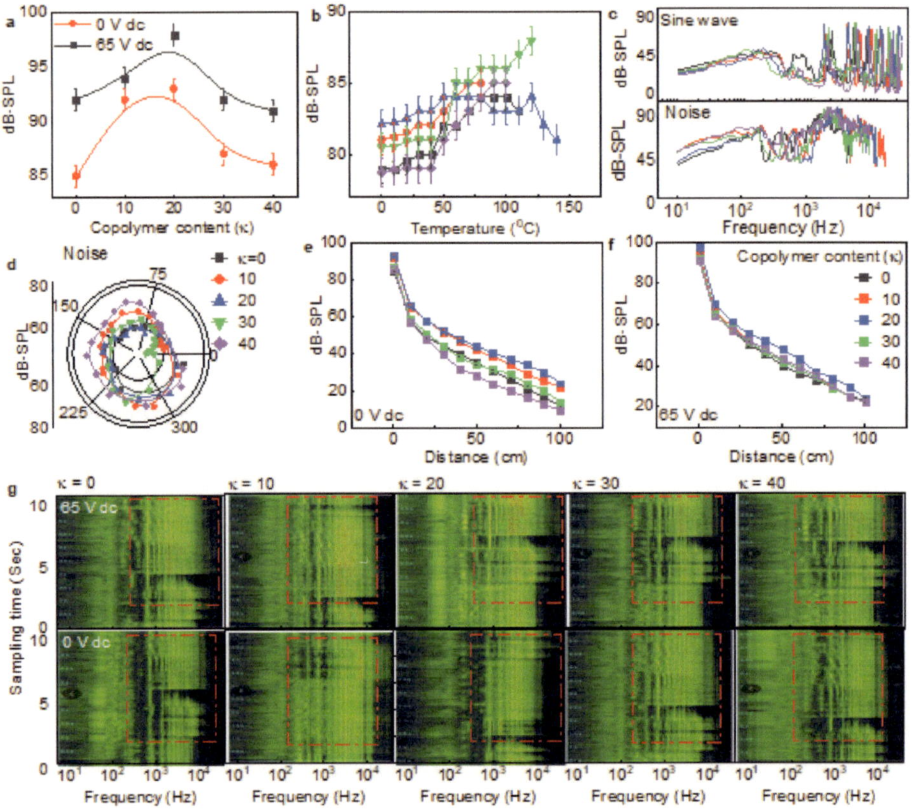

Figure 11.
CTs are made up of polymer blends. (a) SPL graph of blends at V = 0 and 65 V_{dc}. (b) Sound Pressure Level (SPL (dB)) Vs T graph. (c) SPL Vs noise at the graph. (d) Noise polar graph at variable frequencies (Hz). (e-f) Inverse graph of blends at V = 0 and 65 V_{dc}. (g) Frequency spectra graph (SD: n = 5) [79].

sensitivity of 0.9 V/Pa at approximately 10^0 Hz and 0.9 mV/Pa at 10^2 Hz for the Alpha blend. This resulted in clear speech recognition with minimal noise for both Alpha and Beta blends, exhibiting distinct frequency spectra due to their unique PE and AFE properties. In the TIO transducers, Beta CLU provided +5 dB-SPL sound efficiency, while the Beta CMU successfully recognized speech for sound signals ranging from 23.5 to 440 Hz at a voltage of 65 Vdc. The advancements in TIO, CLUs, and CMUs are poised to enhance smart devices by enabling simultaneous sound transmission and recording through a single transducer unit. **Figure 12** shows the information of CLUs and CMUs made up of blend system. **Figure 12(a)** shows the response of blends with respect to their voltage and operational frequency. **Figure 12(b)** represents the sensitivity of the sound intensity in which Gamma blend shows very good SI at 102 Hz. **Figure 12(c-d)** shows the sinewave graphs of the CT. **Figure 12(e)** represents the color visualization spectra of operational frequency for the blend system.

4. Conclusions

The electroactive properties of ferroelectric polymer blends were discussed, highlighting their advantages in terms of pyroelectric, piezoelectric, and

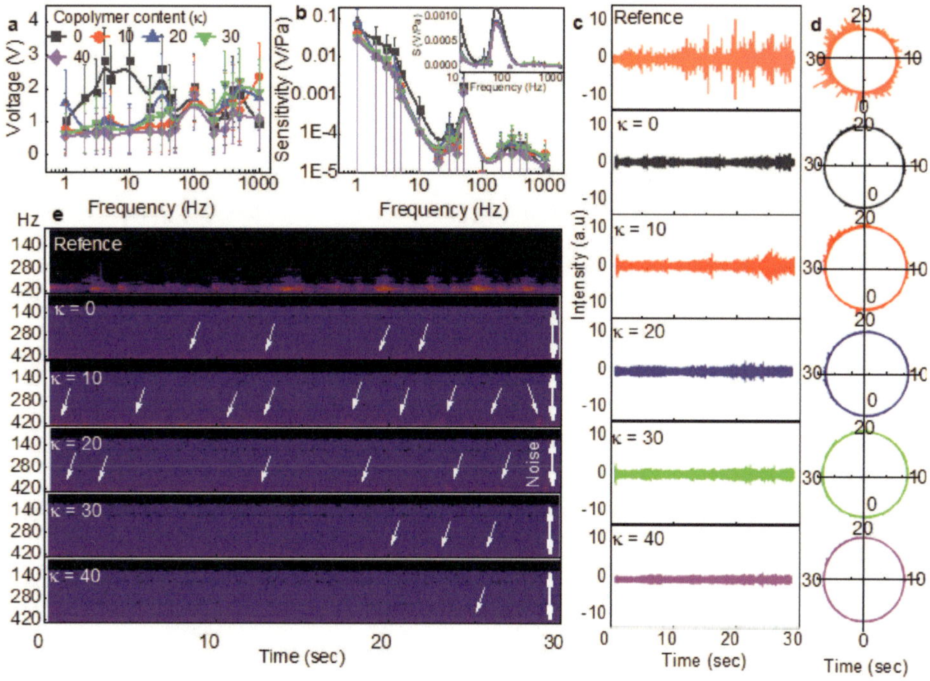

Figure 12.
CTs are fabricated by the polymer blend system. (a) Frequency (Hz) and Voltage trend. (b) Sound intensity (SI) plot. (c,d) Sinewave graphs of transducers. (e) Color visualization spectra of operational frequency for Alpha, Beta, Gamma, Delta, and Exi polymer blend systems (SD: n = 5) [79].

flexoelectric energy harvesting for flexible electronics. Well-known polymers, P(VDF-TrFE) and P(VDF-TrFE-CTFE) were mixed in a specific ratio, resulting in an antiferroelectric mixture through the creation of defect structures in the polymer blends. The antiferroelectric polymer blends were subjected to poling to enhance their beta phase, making them suitable for use as electroactive polymers in flexible electronics.

In this chapter, the pyroelectric, piezoelectric, and flexoelectric properties of the electroactive antiferroelectric polymer blend system are studied. It is reported that the electrostrictive properties of the blend system are enhanced (i.e., percentage of crystallinity enhanced from 10–70%, Energy Density is enhanced up to 1.44 J/cm^3 at near body temperature, the efficiency of Olsen cycle was increased by three fold, SPL is enhanced up to 91 dBSPL for flexible loudspeaker, sensitivity is enhanced by 10^3 mV/pa at 10^2 Hz in flexible microphone and SPL of TIO transducer was increased up to 96 dBSPL by applying the addition dc power source,) by increasing the copolymer content in the terpolymer matrix, which gradually increases its beta phase. This makes the material suitable for fabricating energy-harvesting devices and electro-acoustic transducers for flexible electronics.

Acknowledgements

This work was supported by the National Natural Science Foundation of China (52072178).

Conflict of interest

There is no conflict of interest of all authors.

Notes/thanks/other declarations

We are thankful to Dr. Adeela Nasreen, Dr. Fahad Bin Zahid, and Dr., Zaighum Abbas Naqvi for their support in reviewing the document.

Author details

Mudassar Shehzad[1,2], Liu Yang[1] and Yaojin Wang[1*]

1 School of Materials Science and Engineering, Nanjing University of Science and Technology, Nanjing, China

2 Department of Materials Science and Engineering, Pak-Austria Fachhochschule: Institute of Applied Sciences and Technology, Haripur, Khyber Pakhtunkhwa, Pakistan

*Address all correspondence to: yjwang@njust.edu.cn

References

[1] An N, Liu H, Ding Y, Zhang M, Tang Y. Preparation and electroactive properties of a PVDF/nano-TiO_2 composite film. Applied Surface Science. 2011;**257**:3831-3835. DOI: 10.1016/j.apsusc.2010.12.076

[2] Smith M, Kar-Narayan S. Piezoelectric polymers: Theory, challenges and opportunities. International Materials Reviews. 2022;**67**:65-88. DOI: 10.1080/09506608.2021.1915935

[3] Zhang QM, Bharti V, Zhao X. Giant electrostriction and relaxor ferroelectric behavior in electron-irradiated poly(vinylidene fluoride trifluoroethylene) copolymer. Science. 1998;**280**:2101-2104. DOI: 10.1126/science.280.5372.2101

[4] Casar G, Li X, Koruza J, Zhang Q, Bobnar V. Electrical and thermal properties of vinylidene fluoride–trifluoroethylene-based polymer system with coexisting ferroelectric and relaxor states. Journal of Materials Science. 2013;**48**:7920-7926. DOI: 10.1007/s10853-013-7602-4

[5] Chu B, Zhou X, Ren K, Neese B, Lin M, Wang Q, et al. A dielectric polymer with high electric energy density and fast discharge speed. Science. 2006;**313**:334-336. DOI: 10.1126/science.1127798

[6] Guan F, Wang J, Yang L, Tseng J-K, Han K, Wang Q, et al. Confinement-induced high-field antiferroelectric-like behavior in a poly(vinylidene fluoride-co-trifluoroethylene-co-chlorotrifluoroethylene)-graft-polystyrene graft copolymer. Macromolecules. 2011;**44**:2190-2199. DOI: 10.1021/ma102910v

[7] Zhang Z, Meng Q, Chung TCM. Energy storage study of ferroelectric poly(vinylidene fluoride-trifluoroethylene-chlorotrifluoroethylene) terpolymers. Polymer. 2009;**50**:707-715. DOI: 10.1016/j.polymer.2008.11.005

[8] Su R, Tseng J-K, Lu M-S, Lin M, Fu Q, Zhu L. Ferroelectric behavior in the high temperature paraelectric phase in a poly(vinylidene fluoride-co-trifluoroethylene) random copolymer. Polymer. 2012;**53**:728-739. DOI: 10.1016/j.polymer.2012.01.001

[9] Bharti V, Hu H, Cheng ZY, Mai T, Zhang QM. Quantitative analysis of structural, relaxational and electrostrictive properties of PVDF-TRFE/PMMA films irradiated with high-energy electrons. IEEE Transactions on Dielectrics and Electrical Insulation. 2001;**8**:718-724. DOI: 10.1109/94.946729

[10] Chen X-Z, Li X, Qian X-S, Wu S, Lu S-G, Gu H-M, et al. A polymer blend approach to tailor the ferroelectric responses in P(VDF–TrFE) based copolymers. Polymer. 2013;**54**:2373-2381. DOI: 10.1016/j.polymer.2013.02.041

[11] Kittel C. Theory of antiferroelectric crystals. Physics Review. 1951;**82**:729-732. DOI: 10.1103/PhysRev.82.729

[12] Furukawa T, Takahashi Y, Nakajima T. Recent advances in ferroelectric polymer thin films for memory applications. Current Applied Physics. 2010;**10**:e62-e67. DOI: 10.1016/j.cap.2009.12.015

[13] Kong LB, Huang H, Li S. Fundamentals of ferroelectric materials. In: Ferroelectric Materials for Energy Applications. John

Wiley & Sons, Ltd; 2018. pp. 1-31. DOI: 10.1002/9783527807505.ch1

[14] Koizumi N, Murata Y, Tsunashima H. Polarization reversal and double hysteresis loop in copolymers of vinylidene fluoride and trifluoroethylene. IEEE Transactions on Electrical Insulation. 1986;**EI-21**:543-548. DOI: 10.1109/TEI.1986.348913

[15] Ducharme S, Reece TJ, Othon CM, Rannow RK. Ferroelectric polymer Langmuir-Blodgett films for nonvolatile memory applications. IEEE Transactions on Device and Materials Reliability. 2005;**5**:720-735. DOI: 10.1109/TDMR.2005.860818

[16] Gregorio R Jr. Determination of the α, β, and γ crystalline phases of poly(vinylidene fluoride) films prepared at different conditions. Journal of Applied Polymer Science. 2006;**100**:3272-3279. DOI: 10.1002/app.23137

[17] Yang L, Li X, Allahyarov E, Taylor PL, Zhang QM, Zhu L. Novel polymer ferroelectric behavior via crystal isomorphism and the nanoconfinement effect. Polymer. 2013;**54**:1709-1728. DOI: 10.1016/j.polymer.2013.01.035

[18] Silva JPB, Sekhar KC, Pan H, MacManus-Driscoll JL, Pereira M. Advances in dielectric thin films for energy storage applications, revealing the promise of group IV binary oxides. ACS Energy Letters. 2021;**6**:2208-2217. DOI: 10.1021/acsenergylett.1c00313

[19] Chang C, Tran VH, Wang J, Fuh Y-K, Lin L. Direct-write piezoelectric polymeric nanogenerator with high energy conversion efficiency. Nano Letters. 2010;**10**:726-731. DOI: 10.1021/nl9040719

[20] Mishra S, Unnikrishnan L, Nayak SK, Mohanty S. Advances in piezoelectric polymer composites for energy harvesting applications: A systematic review. Macromolecular Materials and Engineering. 2019;**304**:1800463. DOI: 10.1002/mame.201800463

[21] Kaval WG, Lake RA, Jr RAC. PVDF-TrFE electroactive polymer mechanical-to-electrical energy harvesting experimental bimorph structure. MRS Advances. 2017;**2**:3441-3446. DOI: 10.1557/adv.2017.397

[22] Wang D, Yuan G, Hao G, Wang Y. All-inorganic flexible piezoelectric energy harvester enabled by two-dimensional mica. Nano Energy. 2018;**43**:351-358. DOI: 10.1016/j.nanoen.2017.11.037

[23] Claude J, Lu Y, Li K, Wang Q. Electrical storage in poly(vinylidene fluoride) based ferroelectric polymers: Correlating polymer structure to electrical breakdown strength. Chemistry of Materials. 2008;**20**:2078-2080. DOI: 10.1021/cm800160r

[24] Wang Y, Luo C, Wang S, Chen C, Yuan G, Luo H, et al. Large piezoelectricity in ternary lead-free single crystals. Advanced Electronic Materials. 2020;**6**:1900949. DOI: 10.1002/aelm.201900949

[25] Campanini M, Trassin M, Ederer C, Erni R, Rossell MD. Buried in-plane ferroelectric domains in Fe-doped single-crystalline aurivillius thin films. ACS Applied Electronic Materials. 2019;**1**:1019-1028. DOI: 10.1021/acsaelm.9b00180

[26] Kawai H. The piezoelectricity of poly (vinylidene fluoride). Japanese Journal of Applied Physics. 1969;**8**:975. DOI: 10.1143/JJAP.8.975

[27] Furukawa T. Ferroelectric properties of vinylidene fluoride copolymers.

Phase Transitions. 1989;**18**:143-211. DOI: 10.1080/01411598908206863

[28] Sessler GM. Piezoelectricity in polyvinylidenefluoride. The Journal of the Acoustical Society of America. 1981;**70**:1596-1608. DOI: 10.1121/1.387225

[29] Eberle G, Schmidt H, Eisenmenger W. Piezoelectric polymer electrets. IEEE Transactions on Dielectrics and Electrical Insulation. 1996;**3**:624-646. DOI: 10.1109/94.544185

[30] Wang Y, Wen X, Jia Y, Huang M, Wang F, Zhang X, et al. Piezo-catalysis for nondestructive tooth whitening. Nature Communications. 2020;**11**:1328. DOI: 10.1038/s41467-020-15015-3

[31] Chen X, Luo L, Zeng Z, Jiao J, Shehzad M, Yuan G, et al. Bio-inspired flexible vibration visualization sensor based on piezo-electrochromic effect. Journal of Materiomics. 2020;**6**:643-650. DOI: 10.1016/j.jmat.2020.06.002

[32] Shehzad M, Malik T. Antiferroelectric behavior of P(VDF-TrFE) and P(VDF-TrFE-CTFE) ferroelectric domains for energy harvesting. ACS Applied Energy Materials. 2018;**1**:2832-2840. DOI: 10.1021/acsaem.8b00478

[33] Annapureddy V, Kim M, Palneedi H, Lee H-Y, Choi S-Y, Yoon W-H, et al. Low-loss piezoelectric single-crystal fibers for enhanced magnetic energy harvesting with magnetoelectric composite. Advanced Energy Materials. 2016;**6**:1601244. DOI: 10.1002/aenm.201601244

[34] Priya S, Song H-C, Zhou Y, Varghese R, Chopra A, Kim S-G, et al. A review on piezoelectric energy harvesting: Materials, methods, and circuits. Energy Harvesting and Systems. 2017;**4**:3-39. DOI: 10.1515/ehs-2016-0028

[35] Annapureddy V, Palneedi H, Hwang G-T, Peddigari M, Jeong D-Y, Yoon W-H, et al. Magnetic energy harvesting with magnetoelectrics: An emerging technology for self-powered autonomous systems. Sustainable Energy & Fuels. 2017;**1**:2039-2052. DOI: 10.1039/C7SE00403F

[36] Annapureddy V, Na S-M, Hwang G-T, Kang MG, Sriramdas R, Palneedi H, et al. Exceeding milli-watt powering magneto-mechano-electric generator for standalone-powered electronics. Energy & Environmental Science. 2018;**11**:818-829. DOI: 10.1039/C7EE03429F

[37] Park S, Kim Y, Jung H, Park J-Y, Lee N, Seo Y. Energy harvesting efficiency of piezoelectric polymer film with graphene and metal electrodes. Scientific Reports. 2017;**7**:1-8. DOI: 10.1038/s41598-017-17791-3

[38] Kishore RA, Priya S. A review on low-grade thermal energy harvesting: Materials. Methods and Devices, Materials. 2018;**11**:1433. DOI: 10.3390/ma11081433

[39] Ryu J, Kang J-E, Zhou Y, Choi S-Y, Yoon W-H, Park D-S, et al. Ubiquitous magneto-mechano-electric generator. Energy & Environmental Science. 2015;**8**:2402-2408. DOI: 10.1039/C5EE00414D

[40] Shehzad M, Wang Y. Structural tailing and pyroelectric energy harvesting of P (VDF-TrFE) and P (VDF-TrFE-CTFE) ferroelectric polymer blends. ACS Omega. 2020;**5**(23):13712-13718. DOI: 10.1021/acsomega.0c00871

[41] Hunter SR, Lavrik NV, Mostafa S, Rajic S, Datskos PG. Review of pyroelectric thermal energy harvesting and new MEMs-based resonant energy conversion techniques. In: Energy Harvesting and Storage: Materials,

Devices, and Applications III. SPIE; 2012. pp. 77-90. DOI: 10.1117/12.920978

[42] Goldsmid HJ. Principles of thermoelectric devices. British Journal of Applied Physics. 1960;**11**:209. DOI: 10.1088/0508-3443/11/6/301

[43] Lang SB. Pyroelectricity: From ancient curiosity to modern imaging tool. Physics Today. 2005;**58**:31-36

[44] Thakre A, Kumar A, Song H-C, Jeong D-Y, Ryu J. Pyroelectric energy conversion and its applications flexible energy harvesters and sensors. Sensors. 2019;**19**:2170. DOI: 10.3390/s19092170

[45] Chauhan A, Patel S, Vaish R, Bowen CR. Anti-ferroelectric ceramics for high energy density capacitors. Materials. 2015;**8**:8009-8031. DOI: 10.3390/ma8125439

[46] Fatuzzo E, Kiess H, Nitsche R. Theoretical efficiency of pyroelectric power converters. Journal of Applied Physics. 1966;**37**:510-516. DOI: 10.1063/1.1708205

[47] Chang D-A, Choh Y-H, Hsieh W-F, Lin P, Tseng T-Y. The role of drying-control chemical additives on the preparation of sol-gel derived PLZT thin films. Journal of Materials Science. 1993;**28**:6691-6698. DOI: 10.1007/BF00356416

[48] van der Ziel A. Solar power generation with the pyroelectric effect. Journal of Applied Physics. 1974;**45**:4128-4128. DOI: 10.1063/1.1663926

[49] Mohammed MS, Auner GW, Naik R, Mantese JV, Schubring NW, Micheli AL, et al. Temperature dependence of conventional and effective pyroelectric coefficients for compositionally graded BaxSr1−xTiO_3 films. Journal of Applied Physics. 1998;**84**:3322-3325. DOI: 10.1063/1.368516

[50] Lu SG, Rozic B, Kutnjak Z, Zhang QM, Lu SG, Rozic B, et al. Electrocaloric effect (ECE) in ferroelectric polymer films. In: Ferroelectrics. London, UK: IntechOpen; 2010. DOI: 10.5772/10483

[51] Wang TT, Herbert JM, Glass AM. The Applications of Ferroelectric Polymers. Blackie and Chapman and Hall; 1988. Available from: https://cir.nii.ac.jp/crid/1130282271181595904 [Accessed: October 6, 2024]

[52] Xia F, Cheng Z-Y, Xu HS, Li HF, Zhang QM, Kavarnos GJ, et al. High electromechanical responses in a poly(vinylidene fluoride–trifluoroethylene–chlorofluoroethylene) terpolymer. Advanced Materials. 2002;**14**:1574-1577. DOI: 10.1002/1521- 4095(20021104)14:21<1574::AID-ADMA1574>3.0.CO;2-#

[53] Tripathi AK, van Breemen AJJM, Shen J, Gao Q, Ivan MG, Reimann K, et al. Multilevel information storage in ferroelectric polymer memories. Advanced Materials. 2011;**23**:4146-4151. DOI: 10.1002/adma.201101511

[54] Kasap S. Principles of Electronic Materials and Devices. 3rd ed. USA: McGraw-Hill, Inc.; 2005

[55] Shehzad M, Ahmed S, Siddiqui F, Bukhari SZA, Zahid FB, Akram F, et al. Frequency-dependent ferroelectric heat cycles in polymer blends: Enhancements in electrocaloric performance of P(VDF-TrFE) and P(VDF-TrFE-CTFE). Polymer. 2025;**316**:127881. DOI: 10.1016/j.polymer.2024.127881, ISSN 0032-3861

[56] Borwick J. Loudspeaker and Headphone Handbook. Focal Press; n.d.. Available from: https://sc.panda985.com/extdomains/books.google.com/books/about/Loudspeaker_and_

Headphone_Handbook.html?hl=zh-CN&id=TscrBgAAQBAJ [Accessed October 6, 2024]

[57] Newell P, Holland K. Loudspeakers: For Music Recording and Reproduction. 2nd ed. New York: Routledge; 2018. DOI: 10.4324/9781315149202

[58] Wang Y-T, Hu Y-C, Chen K-R. A flexible polyvinylidene fluoride film-loudspeaker. 中國機械工程學刊. 2015;**36**:59-66

[59] White DR. Wide-range electrostatic loudspeaker with a zero-free polar response. Journal of the Audio Engineering Society. 2009;**57**:822-831

[60] Ohga J. A flat piezoelectric polymer film loudspeaker as a multi-resonance system. Journal of the Acoustical Society of Japan (E). 1983;**4**:113-120. DOI: 10.1250/ast.4.113

[61] Shehzad M, Wang S, Wang Y. Flexible and transparent piezoelectric loudspeaker. Npj Flexible Electronics. 2021;**5**:1-6. DOI: 10.1038/s41528-021-00121-z

[62] Lee CS, Kim JY, Lee DE, Joo J, Wagh BG, Han S, et al. Flexible and transparent organic film speaker by using highly conducting PEDOT/PSS as electrode. Synthetic Metals. 2003;**139**:457-461. DOI: 10.1016/S0379-6779(03)00199-1

[63] Sugimoto T, Ono K, Ando A, Kurozumi K, Hara A, Morita Y, et al. PVDF-driven flexible and transparent loudspeaker. Applied Acoustics. 2009;**70**:1021-1028. DOI: 10.1016/j.apacoust.2009.03.007

[64] Weigold JW, Brosnihan TJ, Bergeron J, Zhang X. A MEMS condenser microphone for consumer applications. In: 19th IEEE International Conference on Micro Electro Mechanical Systems. 2006. pp. 86-89. DOI: 10.1109/MEMSYS.2006.1627742

[65] Zhang Z, Wang L, Lee C. Recent advances in artificial intelligence sensors. Advanced Sensor Research. 2023;**2**:2200072. DOI: 10.1002/adsr.202200072

[66] Xu SC, Man BY, Jiang SZ, Chen CS, Yang C, Liu M, et al. Flexible and transparent graphene-based loudspeakers. Applied Physics Letters. 2013;**102**:151902. DOI: 10.1063/1.4802079

[67] Piezoelectricity. n.d. Available from: https://sc.panda985.com/extdomains/books.google.com/books/about/Piezoelectricity.html?hl=zh-CN&id=8KyawZ92eIEC [Accessed October 6, 2024]

[68] Park S, Guan X, Kim Y, Creighton FXP, Wei E, Kymissis I, et al. PVDF-based piezoelectric microphone for sound detection inside the cochlea: Toward totally implantable cochlear implants. Trends in Hearing. 2018;**22**:2331216518774450. DOI: 10.1177/2331216518774450

[69] Åkerfeldt M, Lund A, Walkenström P. Textile sensing glove with piezoelectric PVDF fibers and printed electrodes of PEDOT: PSS. Textile Research Journal. 2015;**85**(17):1789-1799. DOI: 10.1177/0040517515578333

[70] Choi S, Jiang Z. A novel wearable sensor device with conductive fabric and PVDF film for monitoring cardiorespiratory signals. Sensors and Actuators A: Physical. 2006;**128**:317-326. DOI: 10.1016/j.sna.2006.02.012

[71] Jung YH, Hong SK, Wang HS, Han JH, Pham TX, Park H, et al. Flexible

piezoelectric acoustic sensors and machine learning for speech processing. Advanced Materials. 2020;**32**:1904020. DOI: 10.1002/adma.201904020

[72] Shintaku H, Nakagawa T, Kitagawa D, Tanujaya H, Kawano S, Ito J. Development of piezoelectric acoustic sensor with frequency selectivity for artificial cochlea. Sensors and Actuators A: Physical. 2010;**158**:183-192. DOI: 10.1016/j.sna.2009.12.021

[73] Lee HS, Chung J, Hwang G-T, Jeong CK, Jung Y, Kwak J-H, et al. Flexible inorganic piezoelectric acoustic nanosensors for biomimetic artificial hair cells. Advanced Functional Materials. 2014;**24**:6914-6921. DOI: 10.1002/adfm.201402270

[74] Lang C, Fang J, Shao H, Ding X, Lin T. High-sensitivity acoustic sensors from nanofibre webs. Nature Communications. 2016;**7**:11108. DOI: 10.1038/ncomms11108

[75] Jang J, Lee J, Woo S, Sly DJ, Campbell LJ, Cho J-H, et al. A microelectromechanical system artificial basilar membrane based on a piezoelectric cantilever array and its characterization using an animal model. Scientific Reports. 2015;**5**:1-13. DOI: 10.1038/srep12447

[76] Je CH, Lee J, Yang WS, Kim J, Cho Y-H. A surface-micromachined capacitive microphone with improved sensitivity. Journal of Micromechanics and Microengineering. 2013;**23**:055018. DOI: 10.1088/0960-1317/23/5/055018

[77] Walser S, Siegel C, Winter M, Feiertag G, Loibl M, Leidl A. MEMS microphones with narrow sensitivity distribution. Sensors and Actuators A: Physical. 2016;**247**:663-670. DOI: 10.1016/j.sna.2016.04.051

[78] Ali WR, Prasad M. Piezoelectric MEMS based acoustic sensors: A review. Sensors and Actuators A: Physical. 2020;**301**:111756. DOI: 10.1016/j.sna.2019.111756

[79] Shehzad M, Wang Y. PVDF based piezoelectric condenser loudspeaker and microphone. Sensors and Actuators A: Physical. 2022;**346**:113861. DOI: 10.1016/j.sna.2022.113861

[80] Edelman PE. Condenser loudspeaker with flexible electrodes. Proceedings of the Institute of Radio Engineers. 1931;**19**:256-267. DOI: 10.1109/JRPROC.1931.222311

[81] Miao H, Li C, Wang J. A future of smarter digital health empowered by generative pretrained transformer. Journal of Medical Internet Research. 2023;**25**:e49963. DOI: 10.2196/49963

[82] Dias T, Monaragala R, Soleimani M. Acoustic response of a curved active PVDF-paper/fabric speaker for active noise control of automotive interior noise. Measurement Science and Technology. 2007;**18**:1521. DOI: 10.1088/0957-0233/18/5/043

[83] Guigou C, Fuller CR. Control of aircraft interior broadband noise with foam-PVDF SMART skin. Journal of Sound and Vibration. 1999;**220**:541-557. DOI: 10.1006/jsvi.1998.1972

[84] Han JH, Bae KM, Hong SK, Park H, Kwak J-H, Wang HS, et al. Machine learning-based self-powered acoustic sensor for speaker recognition. Nano Energy. 2018;**53**:658-665. DOI: 10.1016/j.nanoen.2018.09.030

[85] Elfes A. Sonar-based real-world mapping and navigation. IEEE Journal on Robotics and Automation. 1987;**3**:249-265. DOI: 10.1109/JRA.1987.1087096

[86] Wells PNT. Ultrasonic imaging of the human body. Reports on Progress in Physics. 1999;**62**:671. DOI: 10.1088/0034-4885/62/5/201

[87] Hohm D, Hess G. A subminiature condenser microphone with silicon nitride membrane and silicon back plate. The Journal of the Acoustical Society of America. 1989;**85**:476-480. DOI: 10.1121/1.397699

[88] Scheeper PR, Olthuis W, Bergveld P. A silicon condenser microphone with a silicon nitride diaphragm and backplate. Journal of Micromechanics and Microengineering. 1992;**2**:187. DOI: 10.1088/0960-1317/2/3/016

[89] Ganji BA, Nateri MS, Dardel M. Design and modeling of a novel high sensitive MEMS piezoelectric vector hydrophone. Microsystem Technologies. 2018;**24**:2085-2095. DOI: 10.1007/s00542-018-3770-0

Chapter 7

Multiferroic Materials: Synthesis, Properties, and Sintering

Mohsin Saleem, Muhammad Zubair Khan, Gwangseop Lee, Muhammad Saad Ur Rahman, Fatima Rehan and Jung-Hyuk Koh

Abstract

This chapter explores the potential of multiferroic materials, emphasizing their multiple ferroic modes and advanced synthesis techniques. Multiferroics, which exhibit ferroelectricity, and ferromagnetism are highly promising for energy harvesting, sensors, and spintronic device applications. Recent advances in synthesis processes such as solid-state reactions, sol-gel processing, and hydrothermal methods are discussed. Innovative sintering techniques, including microwave sintering and spark plasma sintering, are highlighted for their ability to enhance the microstructure and multiferroic coupling at room temperature. These improvements are crucial for practical applications in non-volatile memory devices, actuators, and energy-efficient sensors. The chapter also considers future trends, focusing on advancements in processing methods and developing materials with optimized multiferroic properties. Emerging applications in energy harvesting and spintronic devices are explored, underscoring the growing importance of multiferroics in modern technology. By examining recent experimental findings and innovative processing techniques, this chapter provides a clear understanding of the role of multiferroics in technological advancements and their potential to drive future innovations in functional materials.

Keywords: ferroelectric, multiferroic, synthesis, sintering, ceramics material

1. Introduction

Ferroic materials are defined by their ability to spontaneously polarize electric dipoles, magnetize magnetic moments, or induce mechanical strains that are invertible by an external application of an electric field, magnetic field, or mechanical stress, respectively [1]. A class of materials known as multiferroics exhibits two or more ferroic features at the same time, including ferromagnetism (magnetic ordering), ferroelasticity (strain or shape change), and ferroelectricity (electric polarization). This coupling between different ferroic properties allows for unique functionalities that are not present in single ferroic materials, making multiferroics highly attractive for advanced applications in electronics, sensors, and energy systems. Multiferroics and multiferroic ceramic composites are rapidly advancing in materials engineering, especially for microelectronics and micromechatronics applications. Growing demands for reliable materials with stable properties drive the search for new multiferroic materials and efficient production methods, including synthesis and sintering techniques, to meet these requirements. Multiferroic materials, such as $BiFeO_3$, $BaNiF_4$, $TbMnO_3$, YIG, and $MnWO_4$, exhibit simultaneous ferroelectric and magnetic orders, making them promising candidates for advanced multifunctional devices. Their unique properties arise from the interplay of structural, magnetic, and electric characteristics, which can be tailored through various synthesis methods along with sintering techniques. Multiferroics like Bismuth ferrite ($BiFeO_3$), which display ferroelectric and antiferromagnetic characteristics at room temp, came into the limelight at the end of the twentieth century and provided new avenues in complex smart applications like sensors and memory. This study focuses on γ-$BaFe_2O_4$ as a promising multiferroic material, highlighting its unique properties and potential applications [2, 3].

Instead of being primarily controlled by Maxwell's electrodynamics, the coupling between electric and magnetic dipoles in multiferroics is related to the Fermi statistics of electrons and virtual charge fluctuations in Mott insulators [4], which give rise to exchange interactions between spins as shown in **Figure 1**. The magnetoelectric connection produces amazing phenomena, like the great sensitivity of electric polarization to external magnetic fields, even if relativistic factors do play a key part in multiferroic activity. Because it permits voltage control of magnetism in multiferroic insulators, this coupling has significant practical implications as it may lower energy dissipation and improve the effectiveness of magnetic devices. Conversely, multiferroic materials exhibit numerous ferroic orders, including simultaneous ferromagnetism and ferroelectricity. In spintronics, where electronic spin is used in conjunction with charge for data processing and storage, this coexistence opens new possibilities. Such interaction of electric and magnetic ordering in multiferroics may induce a magnetoelectric effect, whereby an electric field can control magnetic properties and vice versa. This opens new perspectives for a new family of memory devices with higher functionalities and energy efficiencies [3]. These materials possess a property of spontaneous polarization, which can be controlled by an external field, thus finding wide applications in fabricating memory devices. A hysteresis curve is also exhibited, as indicated in **Figure 2**; since it resembles the ferromagnetic curve, it is easier to understand and handle [3]. Ferroelectric and multiferroic materials are unique, for their properties and preferences in various applications [6, 7]. Polarization in ferroelectric materials denotes the alignment of

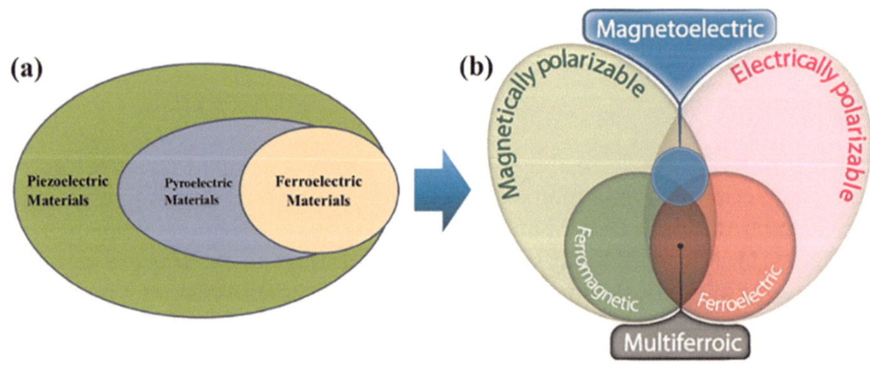

Figure 1.
(a) Relationship between ferroelectric, pyroelectric, and piezoelectric materials and (b) relation between multiferroic and magnetic materials [5].

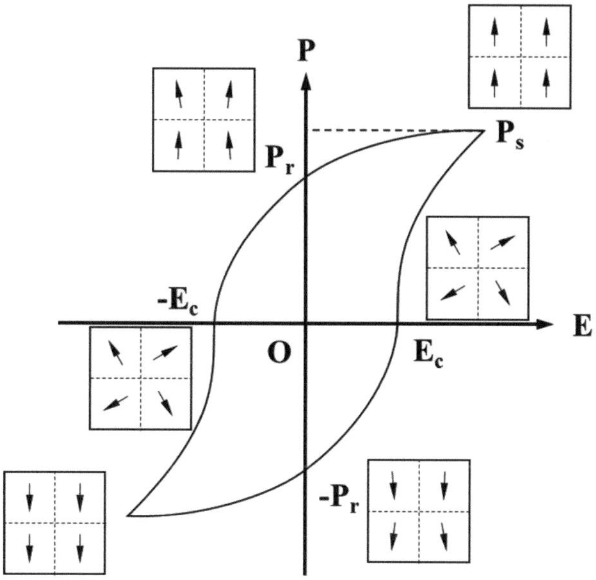

Figure 2.
Typical hysteresis loop illustrating spontaneous polarization (P_s). remnant polarization (P_r), and coercive field (E_c).

electric dipoles resulting from displacements of charged ions inside of the crystal lattice. [1]. This spontaneous polarization occurs on its own, independent of any external electric field, and can be reoriented by the application of an external field. The polarization-electric field—or P-E-hysteresis loop—is often used to depict the polarization behavior of ferroelectric materials. From **Figure 2**, spontaneous polarization represents the organic rotated alignment of dipoles in a ferroelectric material in the absence of an external electric field.

This property differentiates ferroelectric materials from those possessing ordinary dielectric materials. It follows that this asymmetry in the crystal lattice leads to a net dipole moment due to the ion displacement, hence spontaneous polarization. This behavior is temperature-dependent and can change with structural transition. Spontaneous polarization is a key factor in the material's dielectric behavior and its ability to store and retain electric charge. It plays a critical role in applications such as capacitors, where high dielectric constants are desired. Remnant polarization (P_r) refers to the polarization that remains in the material after an external electric field is removed. This "memory" effect is central to the operation of ferroelectric memory devices, where data is stored as polarization states. Such as in Memory Devices (FeRAM), remnant polarization is used to store binary data [8].

1.1 Significant of crystal structure and morphology

In the case of $BiFeO_3$ (BFO), the properties are significantly influenced by variations in particle size, phase composition, grain boundaries, crystallinity, lattice defects, and morphology. These factors directly affect the material's ferroelectric, magnetic, and photocatalytic behavior, including charge separation, polarization, magnetoelectric coupling, and interface activity. Therefore, the choice of synthesis method plays a crucial role in determining the performance of BFO for specific applications. The synthesis techniques for BFO are generally categorized into two approaches: top-down and bottom-up. The top-down approach involves breaking down bulk material into smaller particles through methods such as mechanical milling, laser ablation, and etching. In contrast, the bottom-up approach builds the material through nucleation and growth processes, commonly achieved via solution-based methods like sol-gel synthesis, hydrothermal methods, co-precipitation, combustion, and microwave-assisted synthesis [9]. Each synthesis method influences BFO's microstructure and morphology in unique ways. For instance, sol-gel synthesis often produces fine-grained nanoparticles with high phase purity, while hydrothermal methods can yield nanorods, nanoplates, or nanospheres depending on reaction conditions. Spark plasma sintering (SPS) enhances densification and reduces defects, leading to improved ferroelectric and magnetic properties. Variations in these synthesis parameters can control grain size, crystal orientation, and defect concentration, all of which impact BFO's multiferroic performance and potential applications.

2. Multiferroics materials, structure, and their properties

Multiferroics materials fall into two groups. Multiferroics of Type I: The magnetic and ferroelectric orders exist separately in these materials. Although the link between magnetic and ferroelectric orders is weak, they often show high polarization and ordering temperatures. Type I Multiferroic materials are $BiFeO_3$ [10], $BaNiF_4$, $TbMnO_3$, YIG, and $MnWO_4$ [11]. Type II Multiferroics: In this class, ferroelectricity is induced by magnetic ordering. Stronger magnetoelectric interaction is possible with this kind, where spin order directly causes electric polarization. It would be impossible to include all of the type-II multiferroics that have emerged in the last several years, but some examples include $RFeWO_6$, $LiFe(WO_4)$, $M_2V_2O_7$ [12], Mn_2O_3 [9], Ni_3TeO_6 [13]. A significant challenge with multiferroics is maintaining high coupling efficiency at room temperature, as well as producing materials with reliable and stable properties. To improve these materials' functional qualities, cutting-edge methods like strain engineering and thin-film growth are being investigated [10]. Presently, multiferroics are being investigated for spintronics applications, where electric-magnetic coupling is used for storage. Ferroelectrics are also applied to tunable optical devices because of the electro-optic phenomena [11]. Ferroic materials have become a rediscovery in the field of advanced material sciences to support emerging electronic devices, energy procedures, and communications systems [12]. New fabrication methods like chemical vapor deposition [10] and sol-gel processing have conferred the contextuality of fabricating superior ferroic materials for the purpose of flexible electronics, wearable applications, and smart materials [9]. Multiferroics, in its operating environment, poses complex cross-disciplinary problems of physics, chemistry, and material science. Their coupling mechanisms are crucial for applications in multifunctional device applications, which make ferroelectricity and multiferroicity two of the most critical issues [13].

2.1 Single-phase multiferroics materials and thin film

Single-phase Multiferroics exhibit multiple ferroic orders within a single material, where intrinsic coupling arises from the inherent properties of the material's crystal structure. Single-phase multiferroics hold significant potential for applications in areas such as sensing, memory devices, and quantum control. However, their widespread use has been limited by several challenges. One is related to the low operating temperature since most single-phase multiferroics work well only at temperatures considerably below room temperature. Their electric polarizations in film form are normally very small, further restricting their performance in applications necessitating switchable polarization. Multiferroics capable of working at higher temperatures are mostly weak ferromagnets or display antiferromagnetic behavior, further reducing their practical effectiveness in devices. These factors combined to date have slowed the progress of single-phase multiferroic materials in practical applications, despite their promising capabilities [14].

Recently, various new single-phase multiferroic materials with improved properties have been discovered that guarantee new opportunities for practical applications. Among these, one new frontier is the discovery of a single-phase multiferroic hexaferrite with a giant magneto transport effect at room temperature, representing a major increase in functionality compared to its counterparts. Bismuth Ferrite, $BiFeO_3$, is one of the most-studied single-phase multiferroics; it belongs to the Type I multiferroic material that hosts ferroelectric and antiferromagnetic orders coexisting at room temperature [15]. Bismuth ferrite ($BiFeO_3$) is a notable room-temperature multiferroic material. Its ferroelectricity comes from Bi^{3+} ions, while its magnetic properties arise from Fe ion spins. The material has a ferroelectric transition at 869°C and antiferromagnetic ordering below 369.8°C, with a complex magnetic structure due to spin canting. Interest grew after high polarization ($\sim$60 mC/cm^2) was observed in thin films, surpassing bulk samples limited by oxygen vacancies. Quenching improved bulk polarization to $\sim$20 mC/cm^2, though still lower than thin films. Recent studies [15] show that magnetization in $BiFeO_3$ can be controlled electrically, useful in spin-valve devices [16]. Additionally, researchers have successfully prepared $BiFe_{0.95}Co_{0.05}O_3$ films using a solution-based seeded photocatalytic precursor method, achieving excellent characteristics even at lower preparation temperatures. Furthermore, a single-phase Aurivillius compound has been found to demonstrate intrinsic magnetoelectric coupling at an elevated temperature of 100°C, representing a promising advancement in high temperature multiferroic materials [17]. These breakthroughs suggest that single-phase multiferroics may soon overcome some of their traditional limitations. One of the first multiferroic single-phase thin-film systems studied in research was hexagonal manganite films. Growth mechanisms can be studied using these films as model systems and strain engineering techniques have stabilized phases that would otherwise require bulk high-pressure synthesis. Despite these advancements, their low ordering temperatures and limited polarization values have restricted their practical application [18].

2.2 Composite multiferroics

The term "multiferroic" was initially used to describe single-phase magnetoelectric materials. However, due to their weak coupling between different ferroic orders and the limited magnetoelectric effect at room temperature, their practical applications were restricted, and interest in these materials declined [15]. To address this challenge, numerous studies have been conducted. In 1972, van Suchtelen was the first to propose combining materials with distinct ferroic properties to create composites, resulting in significantly enhanced magnetoelectric effects. The composite multiferroics consist of two or more phases, each contributing unique ferroic properties, with interactions occurring at the interface. Typically, these materials rely on extrinsic coupling mechanisms like strain-mediated interactions, where mechanical strain in one phase influences the ferroic order in the other. This enhances control over the material's properties. Considerable advancements have been made in the creation of composite magnetoelectric systems, which combine electric and magnetic characteristics by means of strain interaction between two materials, usually ferroelectric and ferrimagnetic. Bulk composites were the subject of early research in this area, which revealed experimental magnetoelectric voltage coefficients that did not match theoretical expectations. The possibility of significant magnetoelectric coupling in a multilayer (2–2) configuration was proposed by theoretical models in the 1990s, which made it a desirable option for investigation in thin-film oxide systems [19]. Combining ferroelectric materials such as barium titanate (BT) with ferromagnetic bismuth and cobalt ferrite ($BiFeO_3$ and $CoFeO_3$) can lead to significant magnetoelectric effects. While piezoelectric-ferroelectric coupling is well understood, magnetoelectric coupling is more complex [20].

2.3 Multiferroic 2D materials

Dimensionality plays a crucial role in shaping the properties and physics of materials, making a significant impact on multiferroics as well. Recent studies have focused on investigating 2D magnetism and ferroelectricity, with the discovery of 2D ferromagnetism being a significant breakthrough in condensed matter physics. This has sparked interest in exploring 2D ferroelectricity and multiferroicity. Traditional ferroelectric and multiferroic materials, like perovskites, are typically discussed within a 3D framework, where their large band gaps and low carrier mobility are the focus. In contrast, 2D van der Waals materials offer intriguing possibilities for ferroelectric instability, and the depolarization field effect, a key concern in 3D ferroelectrics, may not be a limitation in true 2D systems. When $BiFeO_3$ is synthesized in 2D form, its multiferroic properties, such as ferroelectricity and antiferromagnetism, can be significantly altered due to the reduced dimensionality and increased surface effects. In 2D BFO, the rhombohedral distortion that drives ferroelectricity in the 3D form may be enhanced or modified, leading to stronger or more easily switchable polarization. The Fe^{3+} ions maintain their magnetic ordering, and the coupling between electric and magnetic orders becomes more pronounced due to the increased influence of surface and interface effects. The multiferroic materials like $BiFeO_3$ (BFO) offer several advantages due to their unique properties and reduced dimensionality. One key benefit is the enhanced magnetoelectric coupling resulting from the close proximity of ferroelectric and magnetic layers, enabling efficient "magnetic reading and electrical writing" functionalities [14]. Unlike 3D systems, 2D materials experience a significantly reduced depolarization field, which allows for stable ferroelectric polarization even at nanometer-scale thicknesses.

In energy harvesting devices, 2D BFO's magnetoelectric coupling can be used to develop efficient vibration energy harvesters at the nanoscale. For spintronics, the ability to control magnetic states via electric fields makes 2D BFO ideal for next-generation spintronic devices. The combined ferroelectric and magnetic properties also make 2D BFO suitable for high-sensitivity sensors and nanoscale actuators. Furthermore, the concept of "magnetic reading and electrical writing" can drive the development of high-density, non-volatile memory devices [21].

2.4 Crystal structure and morphological study of $BiFeO_3$

The unique multiferroic properties of $BiFeO_3$ (BFO) are intrinsically tied to its rhombohedral distorted perovskite crystal structure (space group R3c) as shown in **Figure 3**. This structure consists of Bi^{3+} ions occupying the A-sites, Fe^{3+} ions in the B-sites, and O^{2-} ions forming octahedra around the Fe^{3+} ions. The rhombohedral distortion breaks inversion symmetry, leading to ferroelectric polarization along the [111] direction due to the off centering of Bi^{3+} and Fe^{3+} ions. This non-centrosymmetric structure underlies BFO's piezoelectric effect, where mechanical stress induces lattice distortions and generates an electric charge through ion displacement. Additionally, BFO's crystal structure supports antiferromagnetic ordering of the Fe^{3+} ions, with magnetic interactions mediated by Fe–O–Fe superexchange. The interplay between the ferroelectric and magnetic orders results in the magnetoelectric coupling effect. In this coupling, applying an electric field alters the Fe–O–Fe bond angles, affecting the magnetic order, while applying a magnetic field induces lattice strain that changes the electric polarization. The magnetostrictive effect in BFO, facilitated by the flexible Fe–O octahedra, involves changes in the magnetic domain alignment causing deformation in the lattice. The magnetostrictive effect in $BiFeO_3$ (BFO) arises from the interaction between the magnetic moments of the Fe^{3+} ions and the crystal lattice. When a magnetic field is applied, it alters the alignment of the Fe^{3+} magnetic moments, which, in turn, induces a strain in the crystal lattice, resulting in a physical deformation of the material. This deformation is due to the distortion of the Fe–O octahedra, which directly influences the lattice parameters and enables the magnetostrictive response. The rhombohedral crystal structure of BFO, characterized by its flexible Fe–O bonds, facilitates this strain response, enhancing the material's effectiveness in magnetostrictive applications. This property makes BFO highly suitable for multifunctional devices where magnetic fields can be used to induce mechanical responses, such as in sensors, actuators, and energy conversion systems [23]. The synergy between the piezoelectric and magnetostrictive effects in $BiFeO_3$ (BFO) is essential for its magnetoelectric coupling. When mechanical strain is applied

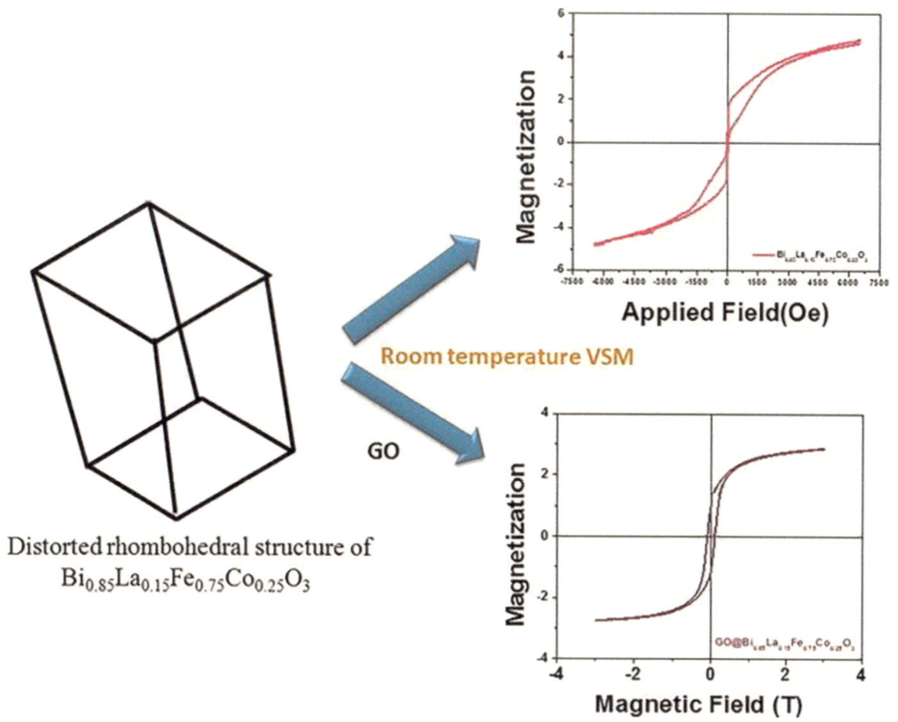

Figure 3.
Crystal structure and multiferroic properties of the BLFCO material [22].

to BFO, it induces a piezoelectric response, generating electric polarization. This strain also modifies the Fe–O–Fe bond angles, affecting the magnetic order and triggering a magnetostrictive response. Conversely, when a magnetic field is applied, it creates lattice strain through magnetostriction, which in turn alters the electric polarization via the piezoelectric effect. This dynamic interplay, enabled by BFO's rhombohedral crystal structure, allows efficient conversion between mechanical, electric, and magnetic energy. As a result, BFO is highly suitable for use in vibration energy harvesters, sensors, and spintronic devices, where such multifunctional properties are critical for performance and efficiency. In the field of energy harvesting, $BiFeO_3$ (BFO) has shown a remarkable photovoltaic response under visible light, making it a promising material for innovative solar cell technologies. Its strong ability to absorb solar radiation and produce charge carriers positions BFO as a potential candidate for advanced heterojunction solar cell designs. Additionally, BFO's inherent piezoelectric and magnetoelectric properties further expand its energy application potential. These properties enable the conversion of mechanical energy into electrical energy in systems subject to strain or vibration and allow for magnetically controlled energy conversion in specialized devices [22].

2.4.1 Effect doping on multiferroic properties of $BiFeO_3$

Doping in $BiFeO_3$ (BFO) significantly enhances its structural, magnetic, and dielectric properties, improving its overall multiferroic behavior. The choice of dopants and the doping strategy, whether single or co-doping, plays a crucial role in optimizing these properties. For instance, cobalt (Co) doping induces a phase transition from rhombohedral to orthorhombic, causing lattice shrinkage and increased strain, which enhances the material's multiferroic performance [24]. Additionally, co-doping with ions like Nd/Sm or Co/Mn introduces greater strain effects due to differences in ionic radii, leading to larger multiferroic properties compared to single-ion doping [25]. In terms of magnetic properties, cobalt-doped BFO exhibits improved ferromagnetism, with saturation magnetization reaching up to 0.956 emu/g [24]. Similarly, doping with Ho^{3+} ions enhances both saturation magnetization and coercivity, reflecting an improvement in magnetic behavior [26]. The dielectric properties also benefit from doping, where interactions between dopants and the Fe^{3+} magnetic sublattice reduce leakage currents and enhance overall stability. For ferroelectric properties, doping with elements such as cerium (Ce) increases remanent polarization, with values peaking at 189.3 $\mu C/cm^2$ depending on concentration [27]. This indicates a significant improvement in ferroelectric behavior. However, while doping generally enhances multiferroic properties, excessive doping can degrade performance, emphasizing the need to optimize doping levels for balanced enhancements across structural, magnetic, and dielectric characteristics.

2.4.2 Effect crystallinity and morphology on multiferroic properties of $BiFeO_3$

The crystallinity and morphology of $BiFeO_3$(BFO) significantly influence its multiferroic properties, which include ferroelectricity and ferromagnetism. Variations in synthesis methods, such as sol-gel and solid-state reactions, lead to different structural characteristics that affect the material's multiferroic performance. The following sections detail how these factors interplay to enhance or diminish multiferroic properties. $BiFeO_3$(BFO) exhibits a distorted rhombohedral structure (R3c space group), which is fundamental to its multiferroic behavior [28]. Structural analysis using X-ray

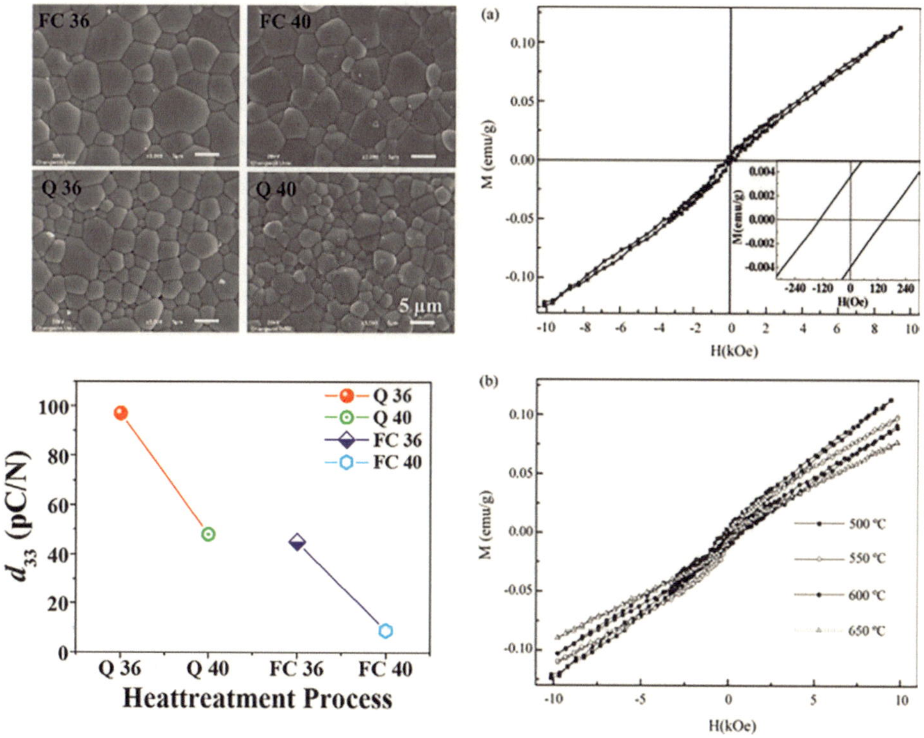

Figure 4.
SEM micrographs of BF–BT furnace cooled and quenched ceramics sintered in conventional furnaces and electromechanical properties at different heating conditions [31] and Magnetic properties of BFO at different temperature [32].

diffraction (XRD) confirms that the degree of crystallinity significantly influences the magnetic and ferroelectric properties. As particle size increases, local strains develop, impacting the overall multiferroic response [29]. The morphology of BFO, particularly its grain size and shape, also plays a critical role in its electrical and magnetic performance. Studies show that nanoparticles exhibit size-dependent ferroelectric coefficients, with smaller particles enhancing ferroelectric properties. Scanning electron microscopy (SEM) reveals that the shape of micrograins influences domain-switching behavior, which is crucial for ferroelasticity [28]. The coexistence of ferromagnetism and ferroelectricity in BFO can be optimized by controlling its crystallinity and morphology, leading to enhanced multiferroic performance. The presence of multidomain structures further modifies these properties, as interactions at domain walls affect polarization and magnetic vectors [30]. However, while improvements in multiferroic properties through controlled crystallinity and morphology are promising, challenges remain in achieving uniformity and scalability in production methods. These challenges may hinder the practical application of BFO in advanced technologies (**Figure 4**).

2.5 Mathematical representation

Multiferroic materials are unique in their ability to exhibit more than one ferroic order simultaneously, such as ferroelectricity, ferromagnetism, and ferroelasticity. This coexistence allows for coupling effects between different order parameters, providing intriguing possibilities for multifunctional devices as already discussed in **Figure 1**. The coupling between electric, magnetic, and elastic order parameters in multiferroic materials enables phenomena such as magnetoelectric and elastoelectric effects [16]. These interactions can be described mathematically, for instance, by the magnetoelectric coupling coefficient α_{ij}, which relates the induced electric polarization P_i to the applied magnetic field H_j through the equation:

$$P_i = \alpha_{ij} H_j \tag{1}$$

Similarly, the elastoelectric effect can be expressed by coupling the mechanical strain S_{ij} with the electric field E_k using an appropriate coupling tensor d_{ijk}. Understanding this coupling mechanism is essential for designing devices that exploit these interactions for applications in memory, sensors, and energy harvesting. Multiferroic is broadly classified into two categories:

$$S_{ij} = d_{ijk} E_k \tag{2}$$

The magnetoelectric effect in multiferroic materials can be described by the following relationships:

$$M = \alpha E^* E \tag{3}$$

$$P = \alpha H^* H \tag{4}$$

Where αE and αH mean electrically and magnetically induced magnetoelectric coupling coefficients. For practical usage, they therefore defined the voltage magnetoelectric coefficient (αH^V). This coefficient relates the voltage response to an external magnetic field:

$$\alpha HV = \partial V/\partial H \tag{5}$$

The Gibbs free energy for a multiferroic system can be expressed as:

$$dG = -SdT + \sigma d\varepsilon - PdE\text{–}MdH \tag{6}$$

The symbol S stands for entropy, ε for strain, P for polarization, and M for magnetization. The constitutive relations derived from this are:

$$P = -\partial G/\partial E, M = -\partial G/\partial H \tag{7}$$

The linear constitutive equations describing the coupling between electric, magnetic, and mechanical fields are:

$$\varepsilon = s\sigma + dE + qH \tag{8}$$

$$P = d\sigma + \chi E\, E + \alpha H \tag{9}$$

$$M = q\sigma + \alpha E + \chi H\, H \tag{10}$$

Where s is the compliance, d and q are piezoelectric and piezomagnetic coefficients, χE and χH are dielectric and magnetic susceptibilities. In composite's multiferroics, the strain-mediated magnetoelectric effect is given by:

$$\alpha HV = \left(\sigma_{\text{magnetic}} \times \sigma_{\text{electric}}\right)/H \tag{11}$$

Where $\sigma_{magnetic}$ stands for the magnetostriction coefficient, and $\sigma_{electric}$ for the piezoelectric coefficient.

2.6 Multiferroic and ferroelectric phase transition

The multiferroic materials are the ABO_3 crystal structures having perovskite ceramics. It is discovered in the single crystal salt form and polycrystalline material in 1921 and 1940 [1]. The appearance of saturation polarization in ferroelectrics is the crystal structure disturbance, that shifted the symmetry of point groups. Ferroelectric has a phase change at Curie point T_c.

$T \sim RT$, $T < T_c$, there is no ferroelectricity in the material.

$T => T_c$, Change of phase near Curie temperature.

Phase transition near Curie temperature, optical, dielectric, thermal, and other properties show unusual performance, due to changes in crystal structure distortion. Dielectric-dependent temperature above T_c in materials is related to the Curie-Weiss Law.

$$\mathcal{E} = C/(T - T_o) \tag{12}$$

Where $\mathcal{E}$ = permittivity of the material, C = Curie constant, T_o = Curie-Weiss temperature.

The first order shows physical change leading to hysteresis during a change of phase. These orders are defined as the finite division in the derivative of Gibbs free energy (G) of the polar phase materials. The nth-order derivative of Gibbs free energy is the finite factor associated with the shift of temperature. Thus, continuous P and S are associated with finite second phase change and infinite factor at first phase order change temperature for ferroelectric materials. The second order is developed by the rapid change of enthalpy, which is most related to physical change [33].

A study on $BiFeO_3$ (BFO) nanoparticles using first-principles calculations reveals that stoichiometric variations, surface termination, and morphology can induce ferromagnetism in this typically antiferromagnetic material. A size-dependent phase diagram shows transitions between ferroelectric, antiferromagnetic, and multiferroic phases as particle size decreases. These findings suggest that tailoring surface and structural effects in nanoscale BFO can unlock enhanced multiferroic properties, offering potential for multifunctional devices in sensors, spintronics, and data storage. The coexistence of ferroelectricity and magnetism in BFO enables electric control of magnetic properties, which is valuable for non-destructive information writing and reading. However, achieving intrinsic multiferroic phases remains challenging due to conflicting requirements for ferroelectricity and magnetism. Innovative strategies, such as nanoscale engineering and stoichiometric control, are crucial for advancing BFO-based multifunctional perovskite oxide devices.

2.7 Different synthesis methods of multiferroic materials

The multiferroic and single-phase multiferroic materials synthesis has significantly benefited from the sophisticated thin-film growth techniques, including pulsed-laser deposition and molecular beam epitaxy. These methods permit total control over material composition and epitaxial strain, thus stabilizing metastable phases and enhancing their multiferroic properties, such as $BiFeO_3$ [34]. The three most common methods include solid-state reaction, the sol-gel method, and hydrothermal synthesis for the synthesis of multiferroic materials.

2.7.1 Solid-state method

The solid-state reaction is quite a general method to prepare ferroic materials. It involves precursor mixing, generally in the form of powders, followed by heating at a high temperature where one can initiate a reaction; the high temperature allows for the diffusion and nucleation of a crystalline phase with the required ferroic properties [35]. In the solid-state route, stoichiometric amounts of Bi_2O_3 and Fe_2O_3 are mixed, milled, and calcined at temperatures ranging from 600 to 1000°C. The reactions are complex due to the formation of secondary phases like $Bi_2Fe_4O_9$ and $Bi_{25}FeO_{39}$, which are challenging to eliminate [36, 37]. The properties of the synthesized $BiFeO_3$ include significant challenges in obtaining a pure single-phase product, but it exhibits both ferroelectric and antiferromagnetic properties, making it suitable for multiferroic applications. However, the synthesis methods influence the presence of impurity phases, affecting the overall magnetic and electric behavior of the material [36].

2.7.2 Sol-gel method

The sol-gel method is another versatile technique, particularly useful for producing highly pure and homogeneous multiferroic materials. In this process, metal oxide alkoxide or metal salts are dissolved in a solvent to form a sol, which then undergoes gelation to form a 3D network. Subsequently, drying and calcination steps yield the desired material. The sol-gel method is advantageous for controlling the material's microstructures and composition at a molecular level, making it suitable for the synthesis of complex oxide materials [13]. The solution undergoes hydrolysis and condensation reactions to form a gel, which is then dried to remove excess solvents. The dried gel is subjected to calcination at temperatures ranging from 500 to 800°C to produce crystalline $BiFeO_3$. The sol-gel process yields nanoparticles with controlled morphology and fine grain sizes, typically in the range of 50–200 nm, depending on processing conditions. One of the key advantages of the sol-gel process for $BiFeO_3$ synthesis is its ability to achieve high phase purity and uniform distribution of Bi and Fe ions at the molecular level. This helps minimize the formation of unwanted secondary phases, which often degrade the material's multiferroic properties. The resulting fine-grained $BiFeO_3$ exhibits enhanced ferroelectric and magnetic properties, such as improved remnant polarization and reduced leakage current, making it suitable for applications in sensors, memory devices, and spintronics.

However, the sol-gel process requires precise control of reaction conditions such as pH, temperature, and precursor concentration. Deviations in these parameters may lead to incomplete reactions or impurities in the final product. Additionally, the drying and calcination steps must be carefully managed to avoid the formation of cracks or agglomerates, which can affect the microstructure and performance of the synthesized $BiFeO_3$.

2.7.3 Hydrothermal/solvothermal method

Hydrothermal method synthesis is employed to produce multiferroic materials under high-pressure and moderate-temperature conditions in an aqueous solution and solvothermal is solvent according to the precursors. This method enables the growth of high-quality single crystal and nanomaterial with controlled morphology and size. Hydrothermal synthesis is particularly effective for creating complex oxides and composite materials. For example, multiferroic bismuth ferrite ($BiFeO_3$) can be synthesized using the hydrothermal method, resulting in a well-defined crystal structure and enhanced functional properties.

In the hydrothermal synthesis method, soluble salts containing bismuth (Bi^{3+}) and iron (Fe^{3+}) ions are combined in a stoichiometric 1:1 molar ratio to form $BiFeO_3$ (BFO). To achieve a homogeneous solution, an organic solvent like ethylene glycol or a stabilizer such as polyvinyl alcohol (PVA) is introduced while stirring continuously. The resulting solution is then transferred to a high-pressure autoclave and heated at temperatures typically ranging from 160 to 230°C for several hours. The temperature, pressure, reaction time, and pH significantly influence the microstructure and phase purity of the synthesized BFO nanoparticles. For instance, lower temperatures around 160°C can produce well-crystallized BFO nanoparticles with average sizes ranging from 10 to 50 nm. Studies have shown that precise control of these conditions helps avoid the formation of secondary phases such as $Bi_2Fe_4O_9$ and $Bi_{2.5}FeO_9$, which are commonly encountered in other synthesis methods.

The hydrothermal method is particularly effective in producing high-purity BFO nanoparticles with controlled morphologies, such as nanorods, nanospheres, and nanoplates. These nanoparticles typically exhibit enhanced ferroelectric and antiferromagnetic properties due to their fine grain size and high crystallinity. For example, hydrothermally synthesized BFO nanoparticles display higher remanent polarization and improved magnetic ordering compared to those produced via conventional solid-state reactions. However, while the hydrothermal method offers advantages like lower synthesis temperatures and superior control over particle morphology, it also has limitations. The need for high-pressure equipment (autoclaves) restricts scalability for bulk production. Additionally, the reaction times can be long, often requiring several hours or even days to achieve the desired crystallinity and phase purity (**Table 1**) [38].

Criteria	Solid-state reaction	Hydrothermal	Sol-Gel
Crystal Lattice Quality	Good, but with defects due to high temperatures	High-quality, well-crystallized lattice	High-quality, fine-grained lattice
Particle Size	Micron-sized particles	Nanoparticles (10–100 nm)	Nanosized particles (50–200 nm)
Phase Purity	Lower; secondary phases are common	High purity with controlled phases	High purity with fewer secondary phases
Processing Temperature	800–1000°C	100–300°C	500–800°C
Processing Time	Several hours	Several hours to days	Several hours to days
Morphology Control	Limited control	Excellent control over shape and size	Good control over morphology
Grain Size	Larger grains (1–10 μm)	Ultra-fine grains (10–100 nm)	Fine grains (50–200 nm)
Equipment Requirements	Basic furnaces, low-cost equipment	High-pressure autoclaves	Basic lab equipment (stirring, drying, furnaces)
Scalability	Suitable for bulk production	Limited scalability	Moderate scalability

Table 1.
Summarizes the key differences and strengths of each synthesis method, helping to determine the best approach based on the desired properties and application of $BiFeO_3$ materials.

3. Sintering techniques for multiferroic and ferroelectric materials

Sintering is a very crucial process in the treatment of ceramic materials, especially multiferroic and ferroelectric materials, involving the heating of compacted powders below their melting point. This process will result in atomic diffusion caused by densification, further enhancing mechanical, electrical, and structural properties. Sintering plays an important role in multiferroic materials because it optimizes their microstructure, allowing grain growth and reducing porosity, thus affecting the electrical and magnetic properties. The primary challenges in producing $BiFeO_3$ ceramics are: (i) obtaining phase-pure material and (ii) achieving sintered densities exceeding 90% of the theoretical value. Various methods have been utilized to synthesize phase-pure $BiFeO_3$.

3.1 Conventional sintering

Conventional sintering of ceramic materials is usually carried out in furnaces under high temperatures, generally between 1000 and 1400°C, depending on the material composition, and for a long time, from hours up to several hours. The goal is to facilitate atomic diffusion, leading to the densification of particles, grain growth, and the removal of residual porosity. In conventional sintering, heating is performed in a furnace over several hours (commonly 4–12 hours) to allow for gradual and uniform material consolidation. The process requires equipment and procedures, making it widely accessible for research and small-scale production. Initial setup costs for conventional furnaces are lower compared to advanced sintering techniques, making it economically feasible for basic applications. These conditions of heating allow the slow development and densification of the grain [39]. However, the disadvantages are that it requires high temperature and long processing time, which possibly result in excessive grain growth and deterioration of material performance.

Conventional sintering may not effectively eliminate all porosity. Consequently, the material typically shows moderate ferroelectric properties with a remnant polarization (P_r) around 20–30 $\mu C/cm^2$ and weak magnetic behavior due to structural defects and porosity [40]. High leakage currents are also common due to non-uniform microstructures and residual porosity. The critical factor for residual porosity can compromise mechanical strength, dielectric properties, and overall reliability of the final product. In addition, higher temperatures during sintering require higher energy input, further reducing the viability of that process for manufacturing on large industrial scales [41]. Achieving uniform grain size and microstructure is challenging due to the slow heating rates and lack of precision in temperature control. Inconsistent microstructure can result in variability in material performance.

3.2 Microwave sintering

Microwave Sintering utilizes electromagnetic radiation to heat materials rapidly and uniformly, reducing sintering times and temperatures. Unlike conventional ways, microwave energy directly couples with the material, resulting in volumetric heating through dielectric and magnetic losses. In microwave sintering, temperatures are typically 100–200°C lower than in conventional sintering, with shorter processing times ranging from minutes to hours, depending on the material and desired properties [42]. Benefits compared to traditional sintering include shorter processing time often by a factor of 2–10 lower sintering temperature, and superior properties of the material, including mechanical strength and electrical performance. In fact, this technique is widely used for synthesizing multiferroic materials, most notably bismuth ferrite, $BiFeO_3$ [31, 43, 44]. The rapid and uniform heating minimizes grain growth and porosity, resulting in materials with improved mechanical, dielectric, and magnetic properties. Also, in optical properties, the energy band gap of $BiFeO_3$ sintered in a microwave furnace makes this material an ideal candidate for solar cell applications [45]. Ji et al. performed a comprehensive study comparing BF-BT ceramics produced through microwave sintering and conventional sintering techniques. Their findings indicated that the microwave-sintered samples exhibited finer grain sizes relative to those obtained through conventional sintering [46]. This sintering method helps retain a finer grain structure and minimizes secondary phases, enhancing the ferroelectric and magnetic properties of BFO. Microwave-sintered BFO can achieve a higher remnant polarization (~40–50 μC/cm^2) and reduced leakage currents due to improved grain uniformity and densification. The rapid heating and cooling cycles also promote better multiferroic coupling, making the material suitable for sensors and memory devices [45]. Additionally, the reduced energy consumption and shorter processing times of microwave sintering contribute to more sustainable manufacturing practices by lowering operational costs and minimizing environmental impact. SEM images show the distribution of metal particles in ceramics with varying concentrations after microwave sintering. Due to the rapid heating and cooling rates of microwave sintering, Ni particles preferentially migrate to the triple points of the grains, indicating insufficient time for sintering and particle diffusion into the grain boundaries (**Figure 5**) [43].

3.3 Spark plasma sintering (SPS)

SPS is an effective method to synthesize high-density ferroelectric ceramics, such as $BiFeO_3$, which has superior piezoelectric properties. The optimization of process parameters such as temperature (usually 200–500°C less than the normal process temperature) and pressure (in a range of 20–100 MPa) results in enhanced material properties achieved by SPS, including a fine microstructure and improved mechanical strength. More importantly, SPS offers minimum sintering time, lower temperatures, and higher control over microstructure, which makes it very appropriate for advanced ferroelectric and multiferroic materials production [48].

SPS achieves near-theoretical density (above 98%) in a short time due to the combined effects of rapid heating and pressure. This reduces porosity, improving mechanical strength and dielectric properties. The rapid heating minimizes grain growth, allowing for the retention of fine microstructures. SPS-produced BFO typically exhibits grain sizes between 100 and 500 nm, enhancing ferroelectric and magnetic performance. The controlled, high-speed sintering process reduces the formation of unwanted secondary phases like $Bi_2Fe_4O_9$, which can degrade BFO's multiferroic properties. SPS can achieve full densification at temperatures between 700 and 800°C, which is 100–200°C lower than conventional sintering, preserving BFO's stoichiometry and reducing energy consumption. These factors contribute to superior multiferroic properties, with remnant polarization values exceeding 50 μC/cm^2 and enhanced magnetic behavior due to reduced defects and porosity. It helps improve the polarization, remnant magnetization, and coercive fields of BFO. The reduced grain size and higher density lead to lower leakage currents and enhanced ferroelectric and antiferromagnetic behavior, making BFO suitable for sensors,

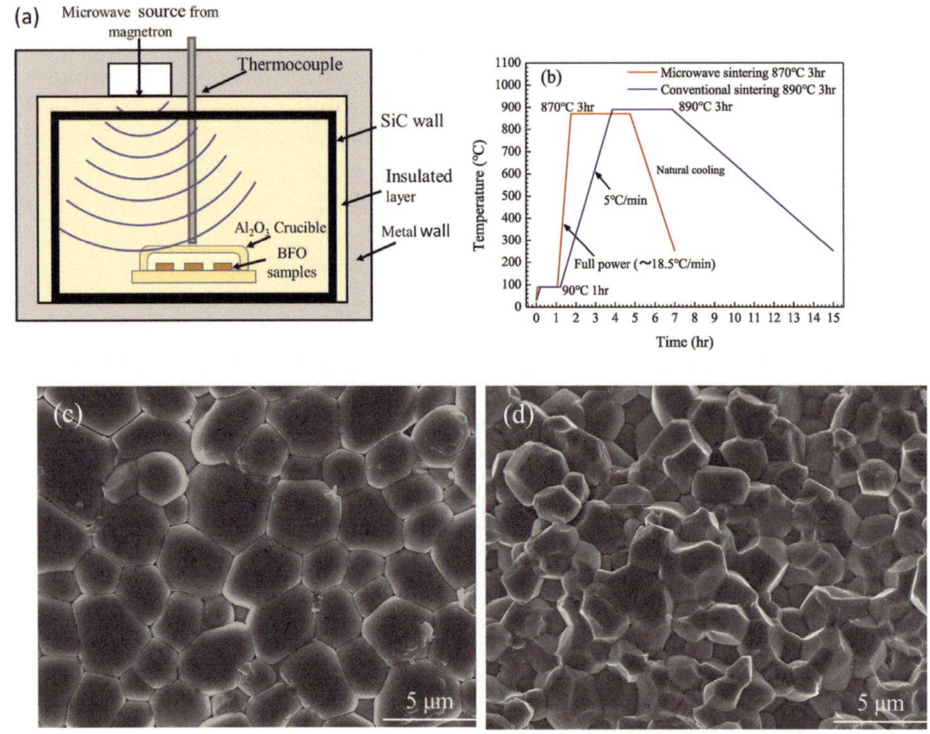

(e)

Atomic (%) (WDS)	Bi	Nd	Fe	O
Stoichiometric ratio	18.6	1.4	20.0	60.0
#1	19.473	1.226	21.159	58.142
#2	19.634	1.273	20.575	58.518
#3	19.697	1.339	20.498	58.466
#4	19.357	1.476	20.88	58.287
#5	19.413	1.347	20.906	58.334
Average	19.515	1.332	20.804	58.349

Figure 5.
(a) Microwave sintering setup for $BiFeO_3$ (BFO) samples. (b) Temperature profiles of microwave (870°C) and conventional sintering (890°C). (c) SEM of microwave-sintered BFO showing uniform grains. (d) SEM of conventionally sintered BFO with larger grains and more porosity. (e) Atomic composition (WDS) of Nd-doped BFO, showing values close to stoichiometric ratios [47].

memory devices, and spintronic applications. Spark Plasma Sintering (SPS) offers rapid densification and fine microstructure control but has drawbacks like high equipment cost, limited scalability, and complex process control. In contrast, conventional sintering is more affordable and scalable for bulk production but suffers from high energy consumption, long processing times, and grain growth. Microwave sintering provides a balance, offering faster processing, lower temperatures, and better microstructure control than conventional sintering, with lower costs and better scalability than SPS. However, SPS outperforms both methods in achieving superior densification and fine grains, making it ideal for specialized, high-performance applications (**Figure 6**).

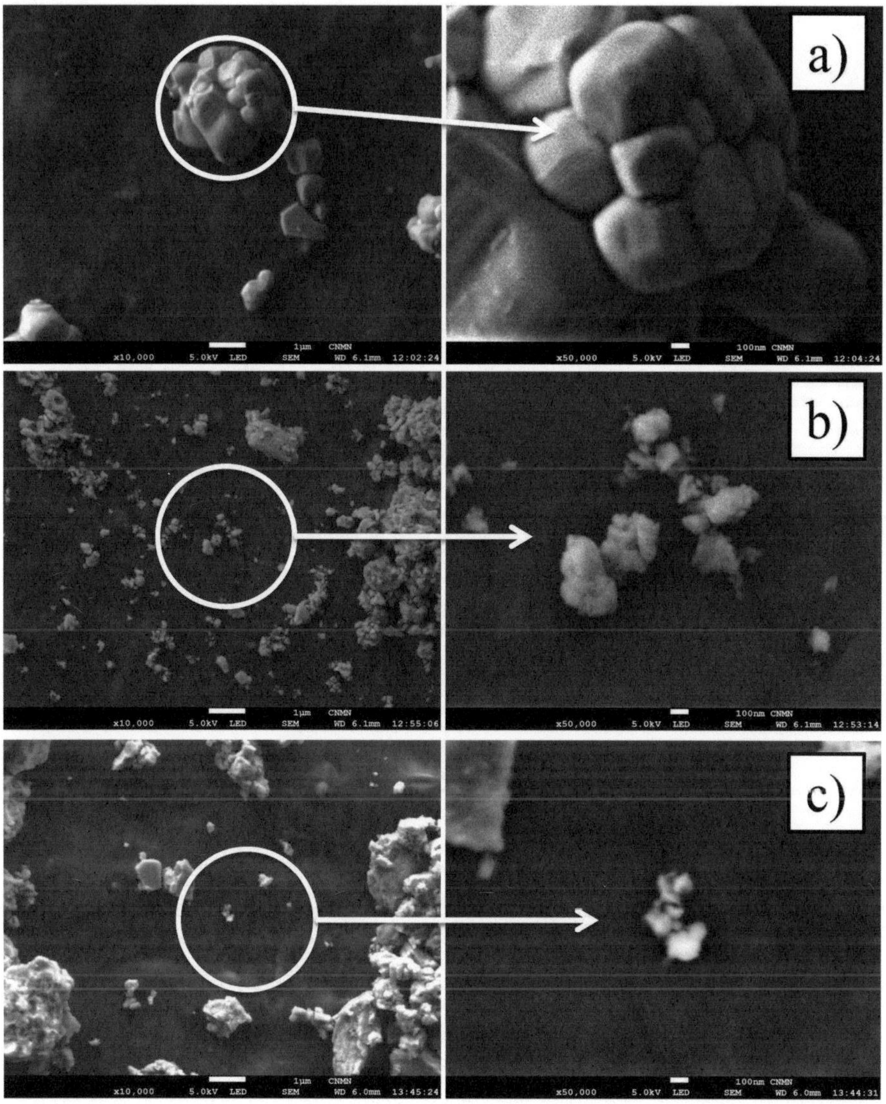

Figure 6.
SEM micrographs of $BiFeO_3$ powders during SPS Sintering at different dwelling times: (a) 0, (b) 30, and (c) 60 min [49].

3.3.1 Interrelationship between properties, structure, and processing

The properties of $BiFeO_3$ (BFO) are intricately linked to its crystal structure and the processing techniques used during synthesis. BFO's crystal structure features a rhombohedral distortion (space group R3c), which plays a fundamental role in its ferroelectricity and magnetoelectric coupling. This structural distortion arises from the displacement of Bi^{3+} ions relative to the FeO_6 octahedra, creating spontaneous polarization and enabling interaction between electric and magnetic orders.

BFO's rhombohedral perovskite structure, characterized by distorted FeO_6 octahedra and displaced Bi^{3+} ions, governs its ferroelectricity, antiferromagnetism, and magnetoelectric coupling. These structural features are sensitive to the choice of

Criteria	Conventional sintering	Microwave sintering	Spark plasma sintering (SPS)
Heating Mechanism	External heating via furnace walls	Internal volumetric heating via microwaves	Joule heating via pulsed DC and pressure
Sintering Temperature	850–1400°C	100–200°C lower than conventional	700–800°C
Processing Time	4–12 hours	Minutes to a few hours	10–20 minutes
Grain Size	Larger grains due to slow heating	Finer grains due to rapid heating	Very fine grains due to rapid densification
Densification	Moderate, higher porosity	Higher densification, lower porosity	Near-theoretical densification
Equipment Cost	Low	Medium	High
Scalability	Good for bulk production	Challenging for large-scale production	Limited to small and medium-sized samples
Energy Efficiency	High energy consumption	More energy-efficient than conventional	Highly energy-efficient
Microstructure Control	Limited control	Better control than conventional	Excellent control
Multiferroic Properties of BFO	Moderate ferroelectricity, higher leakage currents, limited magnetization	Improved ferroelectricity, reduced leakage currents, enhanced magnetic ordering	High polarization, low leakage, enhanced magnetization, and antiferromagnetic properties
Applications	Large-scale production of ceramics, refractories	Functional ceramics, sensors, dielectric materials	High-performance ceramics, sensors, memory devices

Table 2.
Summarizes the key differences and strengths of each sintering method, helping to determine the best approach based on the desired properties and application of $BiFeO_3$ materials.

synthesis methods and sintering processes, which impact the grain size, crystallinity, phase purity, and defect concentration. Processing methods such as sol-gel synthesis, hydrothermal methods, and spark plasma sintering (SPS) significantly impact BFO's microstructure by affecting factors like grain size, density, and defect concentration. Optimizing synthesis and sintering conditions is essential for minimizing defects such as oxygen vacancies and achieving high phase purity in $BiFeO_3$ (BFO). For instance, carefully controlling the calcination temperature during sol-gel synthesis can prevent the formation of secondary phases, while sintering under a controlled atmosphere helps maintain proper stoichiometry and improve crystallinity. These factors, including grain size, density, and microstructural uniformity, are interdependent and directly influence the stability and efficiency of BFO's multiferroic properties, such as its ferroelectricity and magnetic behavior (**Table 2**).

In thin films, techniques like strain engineering can modify the Fe–O–Fe bond angles, which directly affects both ferroelectric and magnetic behavior. Applying strain can optimize magnetoelectric coupling, making BFO more suitable for multifunctional applications. Additionally, the presence of domain walls in nanostructured BFO can influence leakage current and fatigue resistance, which are critical for reliable device performance (**Figure 7**).

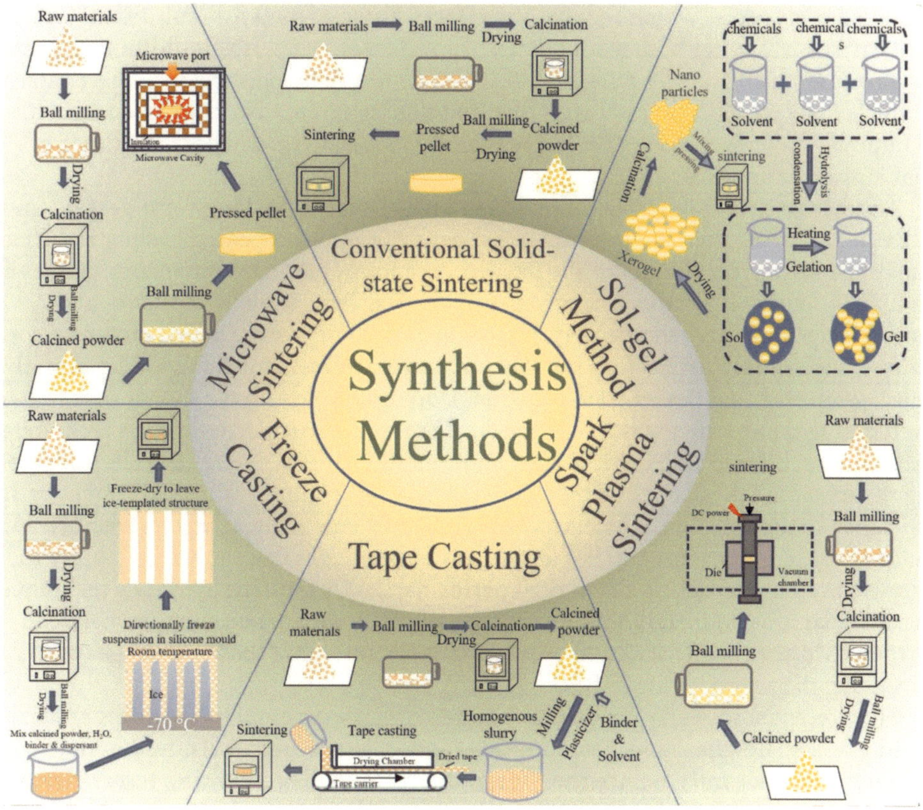

Figure 7.
An overview of ceramic synthesis methods, including solid-state sintering, microwave sintering, freeze casting, tape casting, spark plasma sintering, and the sol-gel method, with step-by-step processes illustrated [50].

By fine-tuning processing conditions such as temperature, pressure, and doping strategies, it is possible to enhance BFO's performance for various applications. These include ferroelectric random-access memory (FeRAM), where high polarization and low leakage current are essential, energy harvesters that convert mechanical energy into electrical energy, and magnetic field sensors that rely on stable magnetoelectric coupling. The interdependence of structure, microstructure, and processing highlights the importance of precise control over synthesis methods to fully exploit BFO's multiferroic potential.

4. Challenges and future perspectives

The multiferroic properties of $BiFeO_3$ (BFO) offer both considerable challenges and promising prospects. While BFO exhibits ferroelectricity and ferromagnetism at room temperature, practical applications are limited by factors such as low coupling efficiency and the presence of antiferromagnetic configurations. BFO's structure, specifically its distorted rhombohedral perovskite (R3c space group), is critical to its multiferroic behavior, with properties influenced by its crystallinity and morphology. Structural analysis via X-ray diffraction (XRD) indicates that local strain develops as particle size. One key challenge is the weak coupling efficiency between ferroelectric and magnetic orders, which reduces the effectiveness of BFO for multifunctional applications, affecting both magnetic and ferroelectric properties. Additionally, the antiferromagnetic order in pure BFO restricts its magnetic performance, necessitating strategies like changing crystal structure and doping or creating composite structures to improve these properties. The current study includes experimental methods of trial-and-error analysis.

Another issue is stability and reusability, which affect the long-term performance of BFO. Achieving consistent multiferroic behavior under repeated cycling is difficult due to the formation of defects like oxygen vacancies and phase impurities. Achieving high phase purity is another challenge, as defects such as oxygen vacancies and secondary phases can degrade performance. Controlling synthesis and sintering conditions, such as calcination temperature and atmosphere, can help optimize crystallinity, grain size, and defect concentration. Doping strategies have shown potential for overcoming these challenges. Morphological control, particularly through techniques like sol-gel synthesis and spark plasma sintering (SPS), plays a crucial role in enhancing the material's properties by reducing defects and refining grain size. Moreover, the efficiency limitations in BFO's performance are influenced by its microstructure and morphological variations. Smaller grain sizes or pore sizes can limit accessibility to perovskite sites, impacting the overall performance. Optimizing synthesis techniques to achieve uniform crystallinity, minimize defects, and enhance coupling between ferroelectric and magnetic orders is essential to overcoming these challenges and advancing the practical use of BFO.

The present study incorporates trial-and-error experimental methods for analyzing multiferroic materials. However, adopting modern modeling and computational techniques can significantly reduce the time and effort required, allowing researchers to design $BiFeO_3$-based multiferroic materials more efficiently using existing databases. These approaches can provide valuable insights to anticipate and resolve potential challenges during the scaling-up process. By leveraging computational analysis, the exploration and development of enhanced multiferroic properties in $BiFeO_3$ become more precise and streamlined, opening up vast opportunities for innovation and advancement in this field.

5. Conclusion

This book chapter reveals a development regarding the presently achieved understanding of multiferroic materials properties, their synthesis methods, and sintering techniques. Such materials, which combine properties like ferroelectricity and ferromagnetism, are opening doors to exciting possibilities from sensors and actuators to memory devices and energy harvesting. The key points that can be reaped include the various synthesis methodologies, such as solid-state reactions, sol-gel routes, and hydrothermal synthesis, used in fabricating such materials with precise control of their characteristic features. Above all, advanced sintering techniques such as conventional sintering, microwave sintering, and spark plasma sintering (SPS) have been mentioned as the key enabling factors for further improvement in microstructure and performance. These techniques refine the material by allowing grain growth, reducing porosity, and enabling optimization in electrical and magnetic properties a necessity for multiferroic effect observation. Considering the outstanding potentialities of multiferroics in perspective devices, one still faces several challenges regarding the appearance of strong coupling of ferroic properties in such materials at room temperature. While these gaps continue to exist, additional studies in the areas of strain engineering, thin-film technologies, and nanostructuring will likely surmount this inadequacy and expand multiferroics' use in advanced technology.

Acknowledgements

This research was supported by the MSIT (Ministry of Science and ICT), Korea, under the ITRC (Information Technology Research Center) support program (IITP-2025-RS-2020-II201655, 100%) supervised by the IITP (Institute of Information and Communications Technology Planning and Evaluation).

Author details

Mohsin Saleem[1,2*], Muhammad Zubair Khan[3], Gwangseop Lee[4], Muhammad Saad Ur Rahman[1], Fatima Rehan[1] and Jung-Hyuk Koh[2,4*]

1 School of Chemical and Materials Engineering, National University of Sciences and Technology, SCME, NUST, Islamabad, Pakistan

2 School of Electrical and Electronic Engineering, Chung-Ang University, Seoul, Republic of Korea

3 Pak-Austria Fachhochschule: Institute of Applied Sciences, Department of Materials Science and Engineering, Haripur, KPK, Pakistan

4 Graduate School of Intelligent Energy and Industry, Chung-Ang University, Seoul, Republic of Korea

*Address all correspondence to: mohsin.saleem@scme.nust.edu.pk and mohsin852@cau.ac.kr and jhkoh@cau.ac.kr

References

[1] Whatmore R. Ferroelectric materials. In: Kasap S, Capper P, editors. Springer Handbook of Electronic and Photonic Materials. Boston, MA: Springer US; 2007. pp. 597-623

[2] Shamsher S, Wasim M, Sabir A, Sahfiq M, Ahmad ARM, Adeefa I. Perspectives of Ferroic/Multiferroic Materials. In: Inamuddin, T. Altalhi and M.A.J. Mazumder, editors. WILEY Online Librairy of Ferroic Materials-Based Technologies. Hoboken, New Jersey: WILEY US; 2024. pp. 35-54

[3] Smolenskii GA, Agranovskaya AI. Dielectric polarization and losses of some complex compounds. Zhurnal Tekhniceskoj Fiziki. 1958;**28**. p. Medium: X; Size: Pages: 1380-1382

[4] Mostovoy M. Multiferroics: Different routes to magnetoelectric coupling. NPJ Spintronics. 2024;**2**(1):18

[5] Pereira N et al. Magnetoelectrics: Three centuries of research heading towards the 4.0 industrial revolution. Materials. 2020;**13**(18):4033

[6] Lines ME, Glass AM. Principles and Applications of Ferroelectrics and Related Materials. Walton Street in Oxford, England: Clarendon Press; 1977

[7] Acosta M et al. Relationship between electromechanical properties and phase diagram in the Ba(Zr0.2Ti0.8)O3–x (Ba0.7Ca0.3)TiO3 lead-free piezoceramic. Acta Materialia. 2014;**80**: 48-55

[8] Arimoto Y, Ishiwara H. Current status of ferroelectric random-access memory. MRS Bulletin. 2004;**29**(11): 823-828

[9] Cong J et al. Spin-induced multiferroicity in the binary perovskite manganite Mn2O3. Nature Communications. 2018;**9**(1):2996

[10] Heron JT et al. Deterministic switching of ferromagnetism at room temperature using an electric field. Nature. 2014;**516**(7531):370-373

[11] Buurma AJC et al. Multiferroic materials: Physics and properties. In: Reference Module in Materials Science and Materials Engineering. 360 Park Avenue South, 6th Floor, New York: Elsevier; 2016

[12] Chen R et al. Magnetic field induced ferroelectricity and half magnetization plateau in polycrystalline R 2 V 2 O 7 (R= Ni, Co). Physical Review B. 2018;**98**(18): 184404

[13] Oh YS et al. Non-hysteretic colossal magnetoelectricity in a collinear antiferromagnet. Nature Communications. 2014;**5**(1):3201

[14] Rondinelli JM, Eidelson AS, Spaldin NA. Non-d 0 Mn-driven ferroelectricity in antiferromagnetic BaMnO 3. Physical Review B—Condensed Matter and Materials Physics. 2009;**79**(20):205119

[15] Zhai K et al. Giant magnetoelectric effects achieved by tuning spin cone symmetry in Y-type hexaferrites. Nature Communications. 2017;**8**(1):519

[16] Vopson MM. Fundamentals of multiferroic materials and their possible applications. Critical Reviews in Solid State and Materials Sciences. 2015;**40**(4): 223-250

[17] Zhao H et al. In: Kimura H, Cheng Z, Jia T, editors. 3 - Aurivillius Layer-Structured Multiferroic Materials, in Nanoscale Ferroelectric-Multiferroic

Materials for Energy Harvesting Applications. Amsterdam, The Netherlands: Elsevier; 2019. pp. 41-60

[18] Fontcuberta J. Multiferroic RMnO3 thin films. Comptes Rendus Physique. 2015;**16**(2):204-226

[19] Martin LW, Chu YH, Ramesh R. Advances in the growth and characterization of magnetic, ferroelectric, and multiferroic oxide thin films. Materials Science and Engineering: R: Reports. 2010;**68**(4):89-133

[20] Wen Y et al. Fabrication and photonic applications of Si-integrated LiNbO3 and BaTiO3 ferroelectric thin films. APL Materials. 2024;**12**(2): 020601-020630

[21] Kanneth Sivaraman S et al. Lead-free piezoelectric energy harvester based on 2D bismuth titanate. ACS Applied Materials and Interfaces. 2023;**15**(49): 57192-57200

[22] Ramam K et al. Magnetic properties of nano-multiferroic materials. Journal of Magnetism and Magnetic Materials. 2017;**442**:453-459

[23] Guo J et al. Strain-induced ferroelectricity and spin-lattice coupling in SrMn O 3 thin films. Physical Review B. 2018;**97**(23):235135

[24] Wang S et al. Enhanced multiferroic properties of Co-doped BiFeO3 nanoparticles via structural changes. Physica Scripta. 2024;**99**(9): 095411

[25] Apostolov AT, Apostolova IN, Wesselinowa JM. Theoretical study of co-doping effects with different ions on the multiferroic properties of BiFeO3 nanoparticles. Materials. 2024;**17**(17): 4298

[26] Gadwala N. Effect of trivalent Ho3+ ion doping on structural, magnetic, optical, and electrical properties of BiFeO3 nanoparticles. Physica Status Solidi (b). 2024;**262**(2):2400304

[27] Lin T et al. Multiferroic properties of BiFeO3 thin films with Ce substitution. Journal of Physics and Chemistry of Solids. 2023;**183**:111662

[28] Carranza-Celis D et al. Control of multiferroic properties in BiFeO3 nanoparticles. Scientific Reports. 2019; **9**(1):3182

[29] Muneeswaran M et al. Structural, electrical, ferroelastic behavior, and multiferroic properties of BiFeO3. Journal of Materials Science: Materials in Electronics. 2020;**31**(16):13141-13149

[30] Wang D et al. Multidomains made of different structural phases in multiferroic BiFeO 3: A first-principles-based study. Physical Review B—Condensed Matter and Materials Physics. 2013;**88**(13):134107

[31] Maqbool A et al. Evolution of ferroelectric and piezoelectric response by heat treatment in pseudocubic BiFeO3–BaTiO3 ceramics. Journal of Electroceramics. 2018;**41**(1):99-104

[32] Jia D-C et al. Structure and multiferroic properties of BiFeO3 powders. Journal of the European Ceramic Society. 2009;**29**(14):3099-3103

[33] Yu Y. Field-Induced Antiferroelectric-Ferroelectric Phase Switching Behavior in Lead Strontium Zirconate Titanate Ceramics. Ann Arbor, Ml, USA: University of Cincinnati; 2000

[34] Martin LW, Schlom DG. Advanced synthesis techniques and routes to new single-phase multiferroics. Current

Opinion in Solid State and Materials Science. 2012;**16**(5):199-215

[35] Prajapati P et al. Effect of sintering temperature on multiferroic properties of Mg-doped bismuth ferrite. Solid State Communications. 2024;**392**:115656

[36] Clabel HJL et al. Insights on the mechanism of solid state reaction between TiO2 and BaCO3 to produce BaTiO3 powders: The role of calcination, milling, and mixing solvent. Ceramics International. 2020;**46**(3):2987-3001

[37] Bernardo MS et al. Reaction pathways in the solid state synthesis of multiferroic BiFeO3. Journal of the European Ceramic Society. 2011;**31**(16): 3047-3053

[38] Thankachan RM, Balakrishnan R. Synthesis strategies of single-phase and composite multiferroic nanostructures. In: Synthesis of Inorganic Nanomaterials. Amsterdam, The Netherlands: Elsevier; 2018. pp. 185-211

[39] Osman KI. Synthesis and Characterization of BaTiO3 Ferroelectric Material. Giza Egypt: Cairo University; 2011

[40] Joshi CR et al. Effect of cobalt substitution on the structural, ferroelectric, and magnetic properties of bismuth ferrite thin films. Journal of Applied Physics. 2022;**132**(19):194102-194113

[41] Wang N et al. Structure, performance, and application of BiFeO3 nanomaterials. Nano-Micro Letters. 2020;**12**:1-23

[42] Gonçalves MD et al. Dielectric characterization of microwave sintered lead zirconate titanate ceramics. Ceramics International. 2016;**42**(13): 14423-14430

[43] Saleem M et al. Dielectric response on microwave sintered BaTiO3 composite with Ni nanopowder and paste. Materials Research Bulletin. 2015; **64**:380-385

[44] Tanvir G et al. Study of ferroelectric and piezoelectric response of heat-treated surfactant-based BaTiO3 nanopowder for high energy capacitors. Materials Science and Engineering: B. 2023;**287**:116100

[45] Harshapriya P, Kaur P, Basandrai D. Influence of La-Ag substitution on structural, magnetic, optical, and microwave absorption properties of BiFeO3 multiferroics. Chinese Journal of Physics. 2023;**84**:119-131

[46] Ji C et al. Influence of sintering method on microstructure, electrical and magnetic properties of BiFeO3–BaTiO3 solid solution ceramics. Materials Today Chemistry. 2021;**20**:100419

[47] Chen P-Y, Chen J-R. Microwave-sintering enhanced photovoltaic conversion in polycrystalline Nd-doped BiFeO3. Ceramics International. 2020; **46**(13):20963-20973

[48] Chavarría CC et al. Low temperature spark plasma sintering of efficient (K, Na) NbO3 ceramics. Journal of the European Ceramic Society. 2024;**44**(6): 4319-4322

[49] Pedro-García F et al. Multiferroic properties of nanostructured BiFeO3 tailored by milling and sintering by SPS. Journal of Alloys and Compounds. 2019; **792**:694-701

[50] Wang B et al. Promising lead-free BiFeO3-BaTiO3 ferroelectric ceramics: Optimization strategies and diverse device applications. Progress in Materials Science. 2024;**146**:101333